Leslie Regan Shade

Gender & Community
IN THE Social Construction
of the Internet

PETER LANG
New York • Washington, D.C./Baltimore • Bern
Frankfurt am Main • Berlin • Brussels • Vienna • Oxford

Library of Congress Cataloging-in-Publication Data

Shade, Leslie Regan.
Gender and community in the social
construction of the Internet / Leslie Regan Shade.
p. cm. — (Digital formations; vol. 1)
Includes bibliographical references and index.
1. Internet and women. 2. Computers and women. 3. Internet—
Social aspects. I. Title. II. Series.
HQ1178 .S52   004′.082—dc21   2001037142
ISBN 0-8204-5023-5
ISSN 1526-3169

Die Deutsche Bibliothek-CIP-Einheitsaufnahme

Shade, Leslie Regan:
Gender and community in the social construction of the Internet / Leslie Regan
Shade. –New York; Washington, D.C./Baltimore; Bern;
Frankfurt am Main; Berlin; Brussels; Vienna; Oxford: Lang.
(Digital formations; Vol. 1)
ISBN 0-8204-5023-5

Cover design by Lisa Dillon

The paper in this book meets the guidelines for permanence and durability
of the Committee on Production Guidelines for Book Longevity
of the Council of Library Resources.

TABLE OF CONTENTS

ACKNOWLEDGMENTS

This book started out as a rather boring thesis in fulfilment of my Ph.D. at McGill University's Graduate Program in Communications. As per many exercises of this sort, it owes a debt of gratitude to several folks at McGill, in particular David Crowley, my supervisor, and Lise Ouimet, departmental assistant par excellence. The Social Science and Humanities Research Council of Canada (SSHRC) was kind enough to grant me a doctoral fellowship in support of this work, and my family was indulgent enough to permit me time amongst the usual domestic rubble. Several months after its completion and interment into the musty bowels of academic-theses-forgotten-but-cataloged, Steve Jones contacted me about resurrecting it for the *Digital Formations* series for Peter Lang. By that time, as with most things on and about the Internet realm, it needed serious updating. I'd like to thank both Steve for his persistence, editor Sophy Craze for her patience, and Ruth Sylvester for her copywriting expertise. A substantial portion of this book was composed in relative beatific splendour at Blue Sea Lake in Quebec, and I thank Ian, Fiona, and Malcolm Duncan for letting me ignore them. Fiona–intrepid and addicted ICQ maniac— deserves thanks for her advice on girl's sites. Thanks also to Ann "separated at birth" Travers for encouraging comments when I was theoretically brain-dead.

Earlier versions of Chapter 5 appeared in *The Information Society* (January-March 1998): 33–44, and Chapter 6 in *Women in Computing*, ed. Rachel Lander and Alison Adam (Intellect Books, 1997).

This book is dedicated to the memory of Lorna Shade, who didn't need the Internet to stay in touch with her many friends and family.

INTRODUCTION

1

Gender, Community, and the Social Construction of the Internet

Despite the belief of some individuals, the computer is not a toy; it is the site of wealth, power, and influence, now and in the future. Women—and indigenous people, and those with few resources—cannot afford to be marginalized or excluded from this new medium. To do so will be to risk becoming the information-poor. It will be to not count; to be locked out of full participation in society in the same way that illiterate people have been disenfranchised in a print world (Spender 1995, xvi).

The feminization of the Internet is a very important shift, because women seek out different Web destinations than men, spend less time surfing online, and are the primary decision-makers in the majority of household purchases, said IDC [International Data Corporation] analyst Frank Gens (Tim Clark 1998).

In the early days of the Internet (say, around 1995 or 1996), way before 24-year-old Marc Andreeson made several million from the IPO of the Netscape World Wide Web browser; way before the Starr Report detailing President Bill Clinton's White House escapades with intern Monica Lewinsky were released on the Internet, causing Web logjams; way before the television show *Oprah Goes Online* debuted on the new women's cable television channel, Oxygen; and way before America Online merged with Time Warner to the tune of $124 billion, the Internet was shockwave-free and participation by women hovered between 15–30% (Shade 1994). Now, as I write this, summer 2000 headlines scream "Wired Women Drive U.S. Internet Group!", "More Women than Men Online in U.S, according to Nielsen Net Ratings", "Majority of Online Shoppers are

Women", and "Girls Turned Off by 'Nerdy' Image of IT" (see NUA Internet Surveys at www.nua.ie).

The Internet is becoming domesticated. It is no longer solely the purview of white, male computer scientists and academics. The current discourse surrounding the Internet positions it as a necessary information and communication tool for the knowledge-based economy of the 21st century. Networking tools provide "Solutions for a Small Planet", as advertising for IBM claims; and a place to go beyond the mundane, according to Microsoft ("Where Do You Want to Go Today?"). Nortel Networks $90 million print and TV advertising campaign, which brands the Beatles song *Come Together* (the rights bought from Sony), asks "What Do You Want the Internet To Be?" and features several heroes and heroines (former astronauts, sports stars of basketball, football and hockey, prominent writers, fashion designers) and regular folks mouthing off pithy statements. [1]

What's more, it is now sexy and politically correct to talk about "bridging the digital divide" by access and participation to the Internet, not just in Organization for Economic Co-Operation and Development (OECD) countries, but in developing countries. The G-8 world leaders that met in Okinawa in July 2000 formed the Dot Force, a "high-level task force dedicated to swooping into developing countries to bridge the 'digital divide' and transport them into the information technology age" (Blanchfield 2000). The Internet is being actively promoted as a tool for eradicating poverty, empowering community, creating jobs, and enhancing democracy. Not everyone agrees with this cyberutopian discourse (Wertheim 1999). Protestors at the G-8 meeting burned a laptop computer on the beach, accusing the leaders of reneging on their promise of $100 billion debt relief for developing countries (the Jubilee campaign). A spokesperson was quoted as saying that "the fundamental cause of the digital divide is poverty" (ibid.) Clearly, such debates over the information society highlight the need for critical reflections on such technological imperatives. If, indeed, cyberspace is a metaphor for community, if it constitutes a network of varied relationships, and if digital citizenship is a prerequisite for participation and engagement in society, then we need to look closely at who is being included, and who is being excluded.

What is fascinating about the Internet now are the tensions between the public and private spheres; and between the commercial versus the nonprofit sectors; and the multifarious definitions of communities; and the debate and negotiations in the global arena about policy on governance, access, and equity. *Gender and Community in the Social Construction of the Internet* is concerned with the creation of women's spaces on the Internet *by and for women*, and also with how women's spaces on the Internet are perceived by corporate interests and policy-makers. "Women" here is defined in general terms, but it is recognized that this is an immense group, with widely differing concerns and perceptions based upon class, race, age, sexual orientation, and ethnic origin. Quite simply, there is no unitary social category for "women". For the most part, the case studies I refer to in this book are based in the North American context, and certainly the gender and cyberspace literature to date is overwhelmingly North American and Western European.

The book has four main objectives. The first is to provide a critical overview, based on the work of diverse feminist scholars, of how certain communication technologies (the telephone, radio, and television) have been historically gendered through the social practices that are promoted and put to use. The second is to provide a very modest overview of some of the diverse women's communities that are using the Internet, especially for issues of feminism, activism, and democracy. The third is to provide a critical feminist and political-economic perspective on the current trajectory of the Internet and its salient qualities of digital capitalism, particularly as women are targeted by corporations as a viable commercial market. The fourth objective is to provide a suggested policy framework on access to the Internet from a feminist perspective, and to provide suggestions for ensuring that women become actively involved in the continued shaping of Internet culture and technology.

RESEARCH ON GENDER AND THE INTERNET

Gender issues surrounding the Internet have been the focus of features in popular journalism, academic discourse and public policy ruminations since the mid-1990s. Much of the mass media coverage has tended to veer towards the hyperbolic (i.e., the Net as a dangerous place for women), or towards the capitalistic imperative (i.e., the Net as a basic entrepreneurial tool for women). In the academic and professional arena, gender issues have been concerned with the participation of women in computer science (Klawe and Leveson 1995; Spender 1995), access (Taylor, Kramarae and Ebben 1993; Light 1995; Balka 1996), social interactions (Cherny and Weise 1996; Winter and Huff 1996; Kendall 2000), the gendering of information systems and design (Green, Owen and Pain 1993; Balka 1997) and issues of sexual harassment and online pornography (MacKinnon 1995; Miller 1995). Research has also focused on gender participation in particular virtual spaces, such as Santa Monica's Public Electronic Network (PEN) (Collins-Jarvis 1993), Amsterdam's Digital City (Rommes, van Oost, and

Oudshoorn 1999) community networks (Travers 2000) and soap opera discussion lists (Scodari 1998; Baym 2000). More recently, work has focused on women participating on the Internet in public spaces, such as cyber-cafes (Lee 1999; Wakeford 1999); and the role of gender and leisure in the domestic context (Green and Adam 1998).

This study differs from other studies on gender and the Internet in that it does not consider the gendered nuances of interpersonal computer-mediated communication (CMC). Linguist Susan Herring, for instance, has analyzed the schematic structure of electronic messages to discover gender differences, and the rhetoric of gender in CMC has been the focus of a special issue of *The Information Society*. [2]

Nor does this book consider how gendered identities are created in Internet communities, such as on Multiple User Dimension Systems (MUDs) and Multi-User Dimension Object Oriented Systems (MOO). Allucquere Stone (1995) and Sherry Turkle (1995), in their respective studies of virtual systems, have been pioneers in this research. Through a description of early and emergent virtual communities and trends in computation, Stone and Turkle have explored the intersection and often conflicting stance between the private and the public spheres that computer-mediated communication, and especially role-playing worlds such as MUDs and MOOs, create. They consider how technologies alter, aid, or constrict women's opportunities for interacting with each other or with the wider public domain. [3]

This book will also not discuss how gender has influenced and impacted the development and cognizance of emerging social mores on the Internet. Several cases, in the North American context, surrounding interpersonal conduct (or netiquette), sexual harassment, privacy, anonymity, identity, and free speech and pornography highlight how gender has been a factor in defining Net norms and developing strategies to resolve, police, or regulate behavior and content on the Internet. [4]

As well, this book does not fit into the technoscience purview, nor does it adopt the cyborgian outlook, famously articulated by Donna Haraway, and adopted by many feminist cyberculture theoreticians. Using the metaphor of the cyborg, these scholars have investigated the relationship between culture and technology, and between gender and machines (Kirkup et al. 2000).

THEORETICAL PERSPECTIVE

What this book does explore are different Internet communities that have been created by women. It is particularly concerned with the creation of

Internet communities by women, which are devoted to issues of feminism, political activism, and the use of the Internet for democratic purposes. It takes up the challenge of Alison Adam, who has written that she is concerned with "what use women *are* making of the new cybertechnologies and which can be used to preserve a sense of political project—even if there is no consensus as to what the politics should be…" (Adam 2000, 286). It is concerned, as well, with what Ann Travers describes as "feminist efforts to re-write the public in cyberspace" (2000, 150).

Women as active agents in the construction of the Internet is a theme woven throughout the book. Mary Beth Haralovich and Lauren Rabinovitz write, "Human agency is a fundamental feminist political goal, a position and contribution for social change as well as an essential aspect of the feminist historical lens" (1999, 8). Their edited collection on American television history illustrates how women, as producers, performers, or fictional characters "are not simply positioned as social victims or passive pawns of oppressive corporations, including the media industries" (ibid.). Although I will be devoting considerable space to looking at the creation of women's communities by women, this book will also chronicle the current emergence and promotion of a commercial Internet environment for women, fueled by the entry of large corporate interests. The social construction of the Internet is characterized not only by the content and objectives of many of the women Internet creators themselves, but also by corporate interests. Both stakeholders alternately subvert and reinforce gender roles. Indeed, the identification of women as a distinct market by for-profit entities, often due to the unanticipated uses of the communication technology by women, is also reflected in the early history of the telephone, radio, and television.

In order to examine these issues, it is necessary to adopt both a feminist social-shaping perspective on technology, as well as a feminist political-economic focus. Although social-shaping or social constructivist ways of looking at technology (MacKenzie and Wajcman 1999) have not directly infiltrated communication studies, they offer a compelling perspective for looking at gender and the Internet. A social-shaping examination of technological systems places an emphasis on the social factors that shape technological change, departing from dominant approaches towards technology that typically study the effects or impact of technology on society. The social constructivist methodology has been soundly criticized by Langdon Winner, when he comments that it has "an almost total disregard for the social consequences of technical choice…what the introduction of new artifacts means for people's sense of self, for the texture of human communities, for qualities of everyday living, and for the broader distribution of power in society…" (Winner 1993, 368). The stance of interpretive flexibility merely reflects "moral and political indifference" (ibid. 372) and an "[un]willingness to examine the underlying patterns that characterise the quality of life in modern technological societies" (ibid. 372).

The social constructivist agenda has also been criticized for not considering gender as constitutive of the relevant social groups in the various case studies of technologies that have typically been examined. The notable exception has been Ruth Schwartz Cowan's examination of the history of home heating and cooking systems in the United States, where she outlined the notion of the "consumption junction". This is "the place and time at which the consumer makes choices between competing technologies" and the place where "technologies begin to reorganize social behavior" (Cowan 1989). By focusing on the consumer, then, the social implications of technologies for individuals and communities can be more easily understood.

Claude Fischer (1992) has also recommended that studies of emergent technologies consider the social uses that individuals make of the technology, the effect on their everyday lives, and the change in social structure as a result of the collective use and response towards a technology. In his social history of the telephone, Fischer has stressed the agency of consumers (notably women and farmers) in adapting the technology of the telephone for myriad social uses, against the predisposed imperatives of the commercial vendors.

By contrast, feminist theories of technology have always been attentive to Winner's insistence that social studies of technology instill a sense of the social consequences of technology. Feminist theories and case studies have been preoccupied with ensuring equitable access to technological know-how in the workplace, in educational settings, and in domestic contexts; with debunking the dominant masculine mythos surrounding technology; and with the creation and practice of environmentally sound communities and technological methods.

As Cynthia Cockburn and Susan Ormrod point out, feminist historical analyses have underscored several conspicuous components missing from mainstream social studies of technology (Cockburn and Ormrod 1993). They point out that a focus on women can highlight the connections between production and consumption, and between production and reproduction. It can also pinpoint the relevant social actors and the gendered assumptions in the design, diffusion, and consumptive stages of a technology's life cycle. An emphasis on the cultural aspects of a technology has been brought center stage, and "these studies show that technological change is quite capable of transforming detailed tasks and activities without changing the fundamental asymmetry and inequality of the relation between women and men" (ibid. 13).

Most importantly, feminist analyses of technology have taken an avowedly political stance, with their ongoing concern with the implications of technologies for women, their work, reproduction, and consumption, and in the wider spheres of the feminine domain: nutrition, horticulture, contraception, childbirth, the environment, and equitable educational and workplace sites (Hopkins 1998).

Some of the trajectories feminist sociotechnical perspectives have taken include: rectifying the historiographical omission of the contributions and participation of women in technological innovation, design, and use; paying attention to technologies that have been ignored or dismissed because they have resided within the women's sphere, such as domestic technologies; examining the historical exclusion of women from the domain of technology, particularly in the labor process; and examining what technologies based on women's values would encompass.

A feminist political-economic focus is also a viable framework to look at issues of gender and the Internet. Feminist communication scholars have traditionally aligned themselves within cultural studies, rather than political-economic perspectives (McLaughlin 1999). This could be because critical communication studies, at least in the United States, have tended to be marginalized within the purview of a select group of universities. Administrative research has been more readily accepted. This tension between functionalist and critical perspectives will probably not be reconciled soon, but it is encouraging to note that many scholars are addressing, in a provocative way, the intersection between communication studies, feminism, and political economy (Meehan and Riordan 2001).

Vincent Mosco, in his seminal study on the political economy of communication, defines political economy "as the study of social relations, particularly the power relations, that mutually constitute the production, distribution and consumption of resources" (Mosco 1996, 25). Four components of political-economy that he outlines include social change and historical transformation; social totality; moral philosophy; and praxis. These components can be readily applied in feminist analyses; certainly praxis, with its emphasis on changing social relations, is apt.

Areas that a feminist political economy of communication should address include the targeting of women as consumers within various media enterprises; the commodification of information; women's access to and participation in the means of production—in this case, communication technologies; the relation between gender and the ownership of media and cultural industries; the impact of gender on global information flows; how the media structures and organizes social relations, with its emphasis on patriarchy and gender socialization; the effect of media and communication technologies on women's work and working conditions; and the international division of labor with respect to technological hardware and software. [5]

Women as a distinct demographic category (or lack thereof) in media industries has been the focus of Eileen Meehan's work (1984, 1990). She has researched how, from the early days of radio, "the demographic category of gender was an industrial concern for the rating monopolist, advertisers, and broadcasters" (Meehan 2000). Thus, the female commodity audience was targeted by daytime talk shows and dramatic serials (soap operas), and advertising was geared towards product placements in the script or commercial interruptions. Meehan has continued to explore the commodity audience, particularly in targeting gendered audiences (Meehan and Consalvo 1999).

The relationship between gender and ownership of media and telecom industries is important in a feminist political-economic context because of the implications of ownership of the media. Women (with the exception of Oprah Winfrey and some female-headed newspaper chain inherited ownership in the U.S.) are simply absent. There are also very few women in the top echelons of new media ("dot.com") firms. [6]

The transnationalization of corporate media and telecom industries has a definite link with gender issues when we think of the international division of labor. This is especially the case with the globalization of work. Offshore production of computer hardware and software, such as computer chips, flow into developing countries (notably India, Malaysia, and the maquilladeros of Mexico) and women are recruited as wage laborers (Sassen 1998). Carla Freeman, in her case study of Caribbean women working in the high-tech industry, argues that "culture and workers' gendered subjectivities must be taken into account in conceptualizing labor markets, labor practices, and the macro-picture of globalization" (2000, 3).

Immigrant women in developed countries are also involved as service and factory production workers. Sometimes this work becomes telework, where the boundaries between public and private life become increasingly blurred (Gunnarsson and Huws 1997). These trends towards the privatization of work, and the blurring between goods and services, creates the "household as virtual workplace" (Menzies 1997, 111). Moreover, the internationalization of call centers, data processing, and software development has led to a new sense of industrial geography. [7]

The social relations and the gendered stereotypes that are mirrored in the mass media and telecom industries are another salient area of research that a critical perspective can shed light on. Understanding the commercial imperatives of the media and cultural industries, which some critics dub a "toxic cultural environment," has been the work of several researchers, most notably Jean Kilbourne (1999). The representation of women in the media (not just in fictionalized accounts and genres, but in news coverage) is the focus of The Global Media Monitoring Project 2000. Over 80 countries monitored the media for its representation of women in newspapers, radio, and television news. Conducted by the World Association for Christian Communication (WACC) and researcher Margaret Gallagher, this is the second global monitoring project to date (the first was conducted in 1995). The 1995 study revealed that while women accounted for 43% of journalists, only 17% appeared as interviewees; and of the interviewees, 29% appeared as victims of accidents or crime. [8]

The palpable tension between the creation of women's spaces for purposes of activism and electronic democracy and the creation of women's communities by corporate and media behemoths who are only concerned with exploiting the commercial potential of women as an audience is evident on the Internet. This is evident with other audiences and uses, as the Internet has become increasingly privatized and commercialized, and as the logic of neoliberalism pervades both the discourse and development of the Internet (Schiller 1999; McChesney 1999). As Michele Martin (2001) emphasizes, it is time to remove the blind spot in feminist studies of new technologies. These studies have been "mostly apolitical and ahistorical and rarely related to wider political, economic or technological elements." According to her, "demystifying the way private capitalist interests have been constructed as the general interest" and asking questions about power are necessary elements for a feminist political-economic perspective.

OVERVIEW OF THE BOOK

Gender and Community in the Social Construction of the Internet is organized into chapters that explore feminist uses of the net, in contrast to efforts by media behemoths and new entrepreneurs to *feminize* the net. By feminization of the Internet, I am specifically referring to the creation of popular content where women's consumption is priviliged and encouraged, rather than production or critical analysis.

Chapter 2, *From the Telephone to the Television: Feminist Perspectives on Communication Technologies,* provides a brief overview of research conducted by feminist communication scholars on the gendering of communication technologies. It looks at how the telephone, radio, and television have been gendered through various social practices widely called acceptable, and also at the unanticipated consequences of women appropriating communication technologies for their own social uses. Themes here include those surrounding consumption and production, the public versus the private spheres, and women as active agents in forging new communities.

It is important to look back at the historical record (or that which has been unearthed so far) in order to become aware that many of the themes and directions with respect to gender and the Internet are not new. This is particularly the case when one looks at the construction of women as negotiators of communication technologies in the domestic arena, and the creation of women as a viable audience and consumer group for advertiser-supported commercial radio and television.

Chapter 3, *Women as Cyberagents*, looks at how various women and women's groups have been actively using the Internet. This includes use of the Internet by political groups for social change (cyberactivism), use of the Internet for cyberfeminist activities, and use of the Internet for support networks, whether the content be academic or health-related. How young women are creatively using the Internet as a means of self-expression and resistance will also be looked at.

The fourth chapter, *Courting Women@E-Com*, chronicles the current emergence and promotion of a commercial Internet environment for women, fueled by the entry of large corporate interests. While it is not surprising that women as a market would be targeted by commercial interests, the tension between e-commerce applications directed towards women as consumers and the usage of the Internet as a locus for citizen-oriented activities provides a useful example of an important issue the Internet faces as digital capitalism continues unheeded.

Chapter 5, *A Gendered Perspective on Access*, examines access issues as they affect women; discusses public policy work on gender equity to national information infrastructure initiatives (now often referred to as the Knowledge Based Economy/Society); and recommends reforms to increase gender equity on the Internet.

Since the 1995 Fourth World Conference on Women, it has become widely accepted that the Internet is crucial for development, and it is feared that those who are unable to access and use information and communication technologies could suffer further marginalization. Several international programs are therefore in place to get developing countries wired. The role of gender in development is crucial, and this has been recognized with the adoption of the Internet.

Chapter 6, *Beyond Beijing: Strategies for the Next Wave,* looks at the issue of women, the Internet, and developing countries, by describing several programs and proposals where women have been actively involved in decision-making regarding the development and implementation of the Internet.

Citizens or Consumers? is the focus of the concluding chapter. What are the democratic possibilities of the Internet for women? What are the key issues with respect to the right to communicate in a media environment characterized by rapid convergence and consolidation? How can the current tension between an industry-defined conception of women online be reconciled with its counterpart, a sisterhood of women online?

Social scientists are now at a key turning point in their explorations of the Internet. A report sponsored by the American Anthropological Society and the Computing Research Association recognizes that advanced information technology will be used by people and institutions in widely varying social contexts. Critical questions that need to be examined include how different groups of individuals and institutions will interact with and be affected by new technologies. By investigating these issues in a broad area of enquiry, technology-driven problems and new opportunities for the uses of technology can be recognized (AAS/CRA 1996).

This book will be of significance to communication scholars interested in the history and theory of gender and communication technologies; social historians of technology; feminist scholars with an interest in a gendered perspective on new communication technologies; and public policy officials interested in how different groups of individuals will interact with, influence, and be affected by new communication and information technologies. The complex issues surrounding community, equity, and the public versus private spheres reflect current and recalcitrant social and policy dilemmas. It is hoped that this book will provide a model for thinking more critically and creatively about our current digital environment.

NOTES

1.   The Nortel Networks ad campaign and its spokesheroes and heroines can be found at www.nortelnetworks.com/corporate/internet/notables/index.html. For a discussion of the infamous MCI "Anthem" television commercial, which proclaimed "There is no race. There is no gender. There is no age. There are no infirmities. There are only minds. Utopia? No, Internet." See Lisa Nakamura, "Where Do You Want to Go Today", pp. 15–26 in *Race in Cyberspace,* ed. Beth Kolko, Lisa Nakamura, and Gilbert B. Rodman (New York: Routledge, 2000).

2.   For Susan Herring's work, see "Gender and Democracy in Computer-Mediated Communication", *Electronic Journal of Communication* (3), edited by Tom Benson, and reprinted in *Computerization and Controversy*, 2nd ed., ed. Rob Kling (New York: Academic Press, 1998); "Bringing Familiar Baggage to New Frontiers: Gender Differences in CMC", plenary speech delivered to the American Library Association, June 27, 1994, and reprinted in *CyberReader*, edited by Victor Vitanza (Boston: Allyn and Bacon, 1998); and Herring's own edited collection, *Computer-Mediated Communication: Linguistic, Social and Cross-Cultural Perspectives* (Amsterdam: John Benjamins Publishing Co., 1996). For an overview of work on gender and CMC, see Herring's "Gender Differences in CMC: Findings and Implications", in the Winter 2000 newsletter of the Computer Professionals for Social Responsibility (CPSR). (URL: www.cpsr.org). "The Rhetoric of Gender in Computer-Mediated Communication" is a special issue of *The Information Society* 15 (1999), edited by Laura J. Gurak and Lisa Ebeltoft-Kraske. Contributors look at issues such as gender harassment, gender dynamics in chatrooms, and avatar design. See also "Sex and Power: Gender Differences in Computer-Mediated Interactions," by N.M. Sussman and D.H. Tyson, *Computers in Human Behavior* 16 (2000): 381–394.

3.   As Turkle and Stone show, communication technologies are gendered through the ways in which they are used to construct us as men or as women (or as other). As well, the social practices with which networked technologies are put to use implicate gendered constructions. Blatant examples of this include online crossdressing and gender-bending role-playing and sexual play (from tinysex to flirtations), to merely a pervasive gendered reticence in using and mastering the computer.

4.   Examples include the "Babes on the Web" case, a creation of a Texan gentleman named Robert Toups. Toups created links to the personal and professional Web pages of a multiplicity of women, and rated the pages using his own ranking system. Some incensed women accused Toups of harassment and electronic stalking, and also of violating intellectual property and copyright norms. See Leslie Regan Shade, "Women, the World Wide Web, and Issues of Privacy," *Feminist Collections* 17 (Winter 1996): 33–35; and Marj Kibby, "Babes on the Web: Sex, Identity, and the Home Page," *Media International Australia* 84 (1997): 39–45. In the Santa Rosa Junior College Case, the U.S. Department of Education's Office of Civil Rights (OCR) initially found that single-sex bulletin boards operated by educational institutions who receive federal financial assistance violate federal sex equity law. The case raises important issues related to: (1) the conflict over sexual discrimination and the First Amendment; (2) debates over single-sex online services, particularly women-only forums; (3) what constitutes sexual harassment online; (4) hate speech codes online, particularly in educational settings; and (5) the ethics surrounding online communication (confidentiality, trust, membership norms). See Leslie Regan Shade, "The Santa Rose Case: Women-Only Forums on the Internet and the First Amendment," *Journal of Information Ethics* (Fall 1997): 48–63. The Mr. Bungle case, as initially reported by Julian Dibbell in *The Village Voice*, became a cause celebre as the issues of whether virtual rape on an MUD could occur became hotly contested and debated both online and in the mainstream media.

See Dibbell's *My Tiny Life: Crime and Passion in a Virtual World* (New York: Owl Books, 1999). The Marty Rimm case (he conducted a study of pornography on the Internet which received sensationalistic coverage in *Time* magazine) not only raised serious criticisms in the Internet community about his methodology, but also fueled the debate over controlling offensive content on the Internet, coming on the heels of the Communication Decency Act (CDA) legislation. See the Cyberporn Debate, a collection of articles and critiques, maintained by Professor Donna Hoffman, at www2000.ogsm.vanderbilt.edu/cyberporn.debate.html.

5.   Thanks to Brandi Bell for her brainpower here.

6.   Exceptions include Carly Fiorina, President and Chief Executive Officer of Hewlett-Packard, and Meg Whitman, President and C.E.O. of eBay. According to *Business Week*, in a poll of Internet executives in the United States, "the executive elite of this new industry looks a lot like the traditional corporate world. Nineteen are white men, four are white women, one is Asian American, and one a Japanese citizen. None is African American or Hispanic." See "The Great Equalizer: Not By a Long Shot," *Business Week* (September 27, 1999).

7.   The EMERGENCE Project (Estimation and Mapping of Employment Relocation in a Global Economy in the New Communications Environment) is carrying out a systematic analysis of the statistics related to teleworking, globalization, and regional development, in order to develop a model for future trends. Core funding is from the European Commission's Information Society Technologies Programme. Ursula Huws is the Project Director. For more information, see www.emergence.nn

8.   Global Media Monitoring Project 2000: www.oneworld.org/wacc/womedia/page3.htm. For the report on the first Global Media Monitoring Project see the MediaWatch site at www.mediawatch.ca/research/gmmp

2 | FROM THE TELEPHONE TO THE TELEVISION

Feminist Perspectives
on Communication Technologies

In the past decade, although feminist analyses of communication technologies have been increasing, the rate at which this research has infiltrated into the discourse and research methodologies of critical communication studies is certainly not as prodigious as the impact of gender and science studies [1]. This could be accounted for partly by the slow integration of feminist thought into the mainstream of communications discourse and by the tardy incorporation of feminist analyses into the prevailing current of technological investigations [2].

Lana Rakow, in explaining how gender has been conceptualized in both the communication field and in communication technologies, writes that "Gender…is usefully conceptualized as a culturally constructed organization of biology and social life into particular ways of doing, thinking, and experiencing the world…gender has meaning, is organized and structured, and takes place as interaction and social practice, all of which are communication processes" (Rakow 1986, 23). To understand the relationships between gender and communication technologies, Rakow advises us not to look for differences in the behavior of men or women towards a technology, but instead to "look for the ways in which the technology is used to construct us as women and men through the social practices that put it to use" (ibid. 23–4).

Three significant implications for understanding the relationship between gender and communication technologies have been explicated by Rakow:

1) Technologies characteristically express male values and meanings; what

is important to notice is how technologies "not only describe but inscribe us" (Rakow 1988a, 67);

2) Technological practices are "constituted...in and through relations of gender. Who does what with a technology for what purposes is, at least in part...an effect of gender. Consequently, not only a technology but also a social practice involving it are associated by gender...Men speak, write, and publish more in the public world of commerce, politics, and ideas (supported by women's labor completing those technological systems), but women write the family letters, make the family telephone calls, and mimeograph the organization newsletter in the private sphere";

3) Technologies are inscribed by gender work' and 'gendered work': "beliefs about the appropriate activities of women and men are acted out in specific social practices, which then seem to confirm the beliefs. By doing the work that is required of their gender women and men are led to experience themselves as gendered, that is, different from each other. Technologies are used to construct and maintain gender differences and hierarchies."

Rakow further argues that more attention needs to be paid to how communication technologies alter, aid, or constrict women's opportunities for interacting with each other, and with the wider public domain. It is also necessary to reiterate the importance of a political economic perspective, paying attention to the often unstated relationship between gender and media ownership, particularly in how women are socially constructed and conceptualized by industry as a viable audience.

This chapter highlights available research on gender and communication technologies, specifically the telephone, the radio, and the television, primarily in the North American context. The impact of communication technologies on the creation and maintenance of community is a theme throughout. Discussion considers the creation of new communities by and for women, and the maintenance and nurturance of existing place-based communities through communication technologies. As well, the development and negotiation of new social norms constitutive of community are apparent in the diffusion of communication technologies.

GENDER AND THE TELEPHONE

The telephone is one of the most ubiquitous communication technologies, yet only in the last two decades has varied research been conducted. This includes sociological and historical examinations, international institutional and policy studies, biographies of the inventors, and even psychoanalytic studies. Not surprisingly, then, the impact of the telephone on women, and their role as both consumers and producers, had, until the mid-1980s, received little consideration from the academic community. [3]

Several recent case studies of the impact of the telephone on gender, and the various methodologies utilized, will be discussed. These include work conducted by Brenda Maddox and Sally Hacker (employment studies); Lana Rakow (ethnographic studies); Carolyn Marvin and Michele Martin (historical research); and Ann Moyal (ethnographic and survey analysis for policy implementation).

EMPLOYMENT STUDIES. Brenda Maddox recognized the importance of examining the role gender played both in women's use of the telephone, particularly as a site of women's labor. In "Women and the Switchboard," she focused on the employment opportunities created for women by the telephone, and the ghettoization of this job as a female occupation. Maddox pointed out how in North America and Europe women were recruited to be operators for the new telephone systems because it was felt that they had the necessary patient temperament, dexterity, and willingness to work for cheap wages that this occupation necessitated. She traced the early history of the operators and the increasing regimentation that the job entailed, and examined contemporary effects of automation on the operators' jobs, and to the sexual discrimination suit launched against AT&T (American Telephone and Telegraph) in the United States during the 1970s (Maddox 1977).

In the mid-1970s, Sally Hacker (1979) conducted a study, "Sex stratification and organizational change," that dealt with the EEOC (Equal Employment Opportunity Commission) investigation into sex and race discrimination at AT&T, then the largest private employer in the United States. In 1972, the EEOC required AT&T to institute affirmative action plans, which would open managerial programs and plant jobs to women and minorities. As a result, 16,300 men were placed in traditionally female jobs but only 9,400 women were placed in traditionally male jobs. In a three-year period, from 1972 to 1975, over 36,000 AT&T jobs were eliminated, most of them operators (due to computerization of the system) and lower-level clerical and managerial staff"—and most of them filled by women.

A decade later, Lana Rakow asked why more research had not been done on the gendered use of the telephone. She broke down the available research into four areas, focusing on geographic and social change; work; universal access and availability; and what she called privatized oppression. Women's dependence on the telephone because of domestic and/or geographic isolation had been noted, but Rakow pointed out that in order to fully understand this relationship, "we must ask why women were isolated and lonely and how the telephone fit into a changing Western landscape of public and private spheres" (Rakow 1988b, 211). Research on women's labor relationship with regard to the telephone could be segmented into paid work as telephone operators, and unpaid telephone labor such as maintaining family and social ties and obligations. What restricts women's access to telephone services? Rakow commented that diffusion studies are complicated, and it is often women, because of their economic situation, who are the last to receive telephone service. Other ways that gender has impacted on accessibility include the privatized oppression of harassing phone calls; the higher rate of single women using unlisted phone numbers or using only their first initials in telephone directories; and the lack of genuine recourse women have when it comes to these offensive and often very threatening experiences.

HISTORICAL RESEARCH. In *When Old Technologies Were New,* Carolyn Marvin studied the formation of a community of experts and technicians around the electronic media of the 19th century—the telegraph, the telephone, and the electric light. Marvin argued that the "early history of electric media is less the evolution of technical efficiencies in communication than a series of areas for negotiating issues crucial to the conduct of social life; among them, who is inside and outside, who may speak, who may not, and who has authority and may be believed" (Marvin 1987, 4). She concentrated not on the technological artifacts themselves, but rather on the way those different social groups interpreted and negotiated the new media. Among the social groups she explored were women.

Marvin focused on the public discourse on and about the new technologies as evidenced in the experts' anecdotes, professional literature, and popular science readings, particularly journals such as *Electrical World, Electrical Review, American Electrician, Scientific American*, and *Telephony*. Through these journals, Marvin was able to survey the emergence of the professional culture of electricians in the late 19th century. These new electrical technicians were the rising priesthood: they invented themselves as an elite and propagated their knowledge through technological chauvinism, which mocked the efforts of outsiders to gain technological literacy. Their targeted groups were ethnic, black, rural, criminal, and female. For this priesthood, intimidation was a means towards preserving their standing, and their creation of a technological canon meant that they controlled the language used to arbitrate the technology.

The electrical experts parodied women's ignorance of the telephone, yet they also encouraged sheltering women from gaining technical knowledge, as they felt that any semblance of women's technological competence would be construed as a threat. In terms of telephone usage, men trivialized the supposed loquaciousness of women's talk, in contrast to their own brisk verbal efficiency.

In *Hello Central*, Michèle Martin integrated feminist analyses of technology, with their emphasis on the class and gender orientation of technology, and political-economic analyses, by looking at the creation of cultural and social practices in the development of the telephone system. Her account offers a fascinating glimpse of the unintended consequences of women's use of a specific technology, and how their use influenced its trajectory. Martin suggests that if women had restricted their telephone usage to the business-oriented imperatives of Bell Telephone, this inconspicuous domestic technology that we take for granted today would not be as ubiquitous as it is now.

Using the rich resources available from the Bell Canada Archives in Montreal, Martin traced the development of the telephone system in Canada, primarily in Ontario and Quebec, from its inception to the system's automation (1876–1920). Utilizing a political-economic perspective, she examined the technical developments of the telephone, from its beginnings as a party-line to its present incarnation as a private-line system. Businessmen were the targeted audience, and Martin relates how Bell Telephone attempted to develop a system suitable for this clientele. Using Montreal as an example, Martin illustrates how the social distribution of telephone service was designed to increase Bell's profits by favoring expansion in wealthy wards. As well, via tempting incentives to subscribe, Bell encouraged the use of the telephone by the professional classes; and concurrently discouraged use of the telephone by the working classes through their reluctance both to install public telephones in convenient locations and to decrease their rates.

Martin is most persuasive when she concentrates on the feminization of the operator labor force, which quickly became a female job ghetto. Bell hired young men who were telegraph operators to assume operator duties, but it was soon discovered that the boys were a fiasco—they were undisciplined, wild, and prone to playing tricks on subscribers. In the wake of the moral fervor and regulation of the late-Victorian period, it was felt that women operators were better able to convey the virtuous qualities dictated by capitalist industry. Therefore, despite the risk and technical knowledge required (electrical know-how because of the telephone's frequent repairs, use of heavy equipment such as a harness head-rest weighing more than six pounds), laborious hours, stuffy conceptions about females in the workforce, and miserably low wages, females were recruited to be operators. The "girls" came from mostly lower to middle and respectable working classes, and applicants were required to adhere to age limits, provide three recommendations (including one from their clergyman), possess a good voice, and not chew gum or wear too many necklaces.

Martin also traced the "making of the perfect operator" (Martin 1991, 50) through Bell's gradual and increasing subjection of the operators to their notion of model behavior. The voice of the operator was used as an ideological instrument of control. Because Bell believed that the voice reflected class origins, working class voices were modified to emulate bourgeois and upper-middle-class voices. As the number of phone calls increased, the speed at which operators placed their calls became a prime concern for Bell. Therefore, new rules of etiquette and brevity of speech were promulgated.

Women subscribers were primarily responsible for developing a viable culture of the telephone, thus appropriating its use in ways unforeseen by Bell, and changing its initial public perception as a "germ collector" and "nerve-racking" technology (ibid. 162). Women tended to use the telephone for socialization, by talking to one another and also for shopping at home, and the rural party-line system allowed for participation in community life by meeting on the lines. This led to the development of the private-line system, where privacy could be better regulated. Such uses compelled Bell to change their developmental strategy to encompass domestic development. The differing notions of the telephone as emancipatory (extending access to the outside world) and repressive (reinforcing sexist ideas of women as aimless gossips) and the contradictory aspects of the privatization and socialization of women's communication are well documented by Martin.

ETHNOGRAPHY: LANA RAKOW AND THE WOMEN OF PROSPECT. Gender on the Line recounts Rakow's 1985 ethnographic field study of women's use of the telephone in a small mid-Western American community she named Prospect. In this work she explored the multifaceted role the telephone assumed in her informants' lives, and argued that the telephone functions, not as a neutral technology, but rather as a gendered technology: "The telephone is a site on which the meanings of gender are expressed and practiced. Use of the telephone by women is both gendered work—work delegated to women—and gender work—work that confirms the community's beliefs about what are women's natural tendencies and abilities" (Rakow 1992, 33).

Rakow examined both the historical and current information about the community and its members, and the impact of Prospect's small telephone company on the community. She conducted interviews with several women participants, presenting the full interviews with the six participants, whose ages and backgrounds varied, "so that their stories, read as a whole, can provide a richer understanding of individual women's lives" (ibid. 9). Rakow entered her research conscious of one of the intrinsic ethical perplexities feminists confront when conducting fieldwork on non-feminist subjects: "...as a feminist I worried about my responsibility to these women and about my ability to resolve the conflict between our interpretations of the world" (ibid. 8). [4]

Investigating the form that women's talk assumes, she argued that it "holds together the fabric of the community, building and maintaining relationships and accomplishing important community functions," as well as fulfilling "important personal needs for individual women as well" (ibid. 34). Such needs emanate from the women's social and economic position in the community, and are met—or not—within the private sphere, which in turn is circumscribed by "economics, technology, and social conventions and sanctions." The women of Prospect were aware that their talk was gender-specific and a result of their social location. Their talk, which they characterized as "visiting" or "gossiping," was limited to domestic concerns and relationships within the private, not the public (and usually masculine) sphere.

Rakow's account of women and the telephone is only one among many that can be told. Prospect was a small, rural, predominantly white, Christian, American community where the women were relegated to a life of limited opportunities imposed by their husband's or father's situation. The telephone assumed an important function in their lives: it was a way of maintaining long-distance communications with family and friends, and a way of easing domestic isolation.

Rakow warns us of the importance of contextualizing studies of gender and technology to understand "the meanings of gender held by the women or group participating, the historical changes that have occurred in social practices and experiences of gender, and the current conditions of women's and men's lives" (ibid. 154). She urges further research into the impact of the telephone on diverse women; for instance, women of color, Jewish women, and educationally privileged women, as well as examining the utilization of the telephone for women's political and economic networking.

POLICY IMPLEMENTATION: ANN MOYAL. Ann Moyal, at the Communications Research Institute of Australia in Canberra, reported on a national Australian study she managed in the late 1980s, whose aims were to survey women's telephone use. The study originated when the Australian Federal Government was going through a period of prospective competitive change in their telecommunications environment, and they wished to ascertain whether the introduction of timed local calls would effect women differently than men.

Moyal's research strategy was exhaustively deep on a small sample. She employed 15 women assistants, with varied personal and professional backgrounds, scattered throughout Australia. In all, the research sample encompassed 200 women aged 16 to 87, from a variety of different socioeconomic and geographical backgrounds. Questions covered demographic data and respondents were asked to complete weekly logs of their telephone usage, from instrumental uses (functional nature: appointments, business dealing, household duties, emergencies, etc.) to intrinsic calls (personal communication with relatives and friends, volunteer work, intimate calls, etc.).

Moyal's survey concluded that there were distinctive feminine patterns of telephone usage. Most of the calls were local (in Australia, this consisted of flat-rate or untimed calls). On average, 2–6 instrumental calls were made per week, with up to 10–12 for women who worked out of their home or were isolated geographically from family and friends. Intrinsic calls were more frequent: on average 14–40 personal calls per week, with 20–28 in metropolitan areas and 14–25 in rural areas. These calls averaged 15–20 minutes and up to 30–45 minutes.

Women's use of the telephone in this study was primarily for "kin-keeping" (maintaining family ties). They also used the phone to maintain friendships. Senior women used the telephone for creating a stable, instant connection for their well-being and health. Telecommuting women and migrant women depended on the telephone. Moyal concluded that "the telephone neighborhood has thus become a key environment for women in Australia…the study has revealed a pervasive, deeply rooted, dynamic feminine culture of the telephone in which kin-keeping, caring, mutual support, friendship, volunteer and community activity play a central part and which, through its ongoing and widening functioning, contributes substantially to women's sense of autonomy, security, participation, and well-being" (Moyal 1992, 67).

In terms of changing from a flat-rate to a metered pay system, most women vociferously agreed that such a practice would detrimentally affect their telephone use. It was recommended to Telecom Australia that timed local calls should not be introduced. Due to market pressures Telecom Australia introduced two packages, one offering flat-rate charges and high equipment rentals; and the other, lower equipment rentals and timed local calling.

FROM REMOTE MOTHERING TO FANTASY FACTORIES. Rakow continued her studies into women's use of the telephone by concentrating on different practices due to technological advancement; specifically, the impact of the cellular phone on mothering. Written as part of a larger research report for UNESCO, *Remote Mothering and the Parallel Shift* investigated the way gender implicated itself into the cellular telephone. Parallel shift referred to mothers who worked outside of the home. She interviewed a group of suburban Chicago women on their use of the cellular telephone, and concluded that men and women used this technology differently. These differences "preserve women's subordinate social position" which are likely to "reproduce gender inequalities, albeit with some shifting of public and private ground, under the guise of solving those very inequities" (Rakow and Navarro 1993, 145).

Rakow characterized women's use of cellular telephones as being distinguished by two tracks—remote mothering and the parallel shift. Although men were originally the primary beneficiaries of cellular phone marketing, women soon became an increasing target. Most of the advertising strategies centered on the need for "supermom" to have access to cellular telephones for both her work and her private, domestic life. Rakow's small survey of suburban cellular telephone users revealed that women were far more likely than their spouses to use cellular telephones for domestic and daily social interactions, such as maintaining the family schedule, keeping track of children's whereabouts, husband's schedule, etc.—what she dubbed "remote mothering."

The lives of women phone sex workers was the subject of Amy Flower's ethnography. As well as interviewing phone sex workers on their perceptions of their work and its effects on their lives, Flood went undercover as a participant-observer in this service industry. What she discovered was the "disembodiment of intimacy" that phone sex affords to the callers, who are able to indulge in diverse sexual fantasies (Flowers, 1998).

GENDER AND RADIO

Unlike the myriad accounts of the gendering of the telephone, the gendering of radio is only now achieving attention in the communication literature. Hilmes contends that the history of women in radio has been "suppressed" (Hilmes 1997, xx), thus constituting yet another blind spot of gender in broadcasting. Therefore, it is necessary to piece together diverse accounts, both historical and contemporary, as to the gendered construction of the radio. Four themes emerge: women as radio amateurs; women as announcers and content innovators; the daytime network schedule as a feminine ghetto to target women as audiences and consumers; and the use of radio during WWII to mobilize women towards the war effort.

Susan Douglas's history of the social construction of early radio in the United States (1899–1922) described how journalistic accounts influenced attitudes towards radio culture, and who should control and use the technology. Douglas commented that "this was primarily a white, middle-class, male construction, a process from which most women and minorities were excluded" (Douglas 1987, xxix). For instance, the press and popular magazines of the day widely reported on early amateur radio operators and inventor-heroes. Amateur (or ham) radio operators were a distinct subculture of males (akin to contemporary computer hackers) who found, through their tinkering and technical mastery of a new technology, a way to cope (and subvert) the pressures of modern bureaucratic conformity. Douglas documented how popular culture fawned over the ideal of the boy inventor-hero: [5]

> Everything could be achieved through technical mastery. Playing with technology was, more than ever, glorified as a young man's game…Few inventions were more accessible to the young man than the latest marvel, wireless telegraphy. Just as articles giving instructions on 'Building Your Own Wireless Set' began appearing with increasing frequency, so did stories and books that celebrated boy wireless experimenters (ibid. 191).

The boy inventor-heroes were able to create their own community of like-minded users:

> In a culture that was becoming more urbanized, and whose social networks were becoming increasingly fragmented, many strangers became friends through wireless. The amateurs loved the contact with invisible others in a realm where one was faceless and yet known at the same time. The fraternity that emerged possessed the fellowship felt among pioneers. These young men were exploring and comparing their friendships on a relatively uncharted and mysterious territory (ibid. 204–5).

However, this community of young men and their dispersed associations were fairly homogeneous: working class boys, immigrants, blacks, and girls were typically not involved in wireless tinkering. Hilmes, however, documents how women, despite social obstacles, were involved as wireless amateurs, through wireless clubs, as Girl Scout activities, and as on-air hams trained by the U.S. Navy during WWI (Hilmes 1997, 132–36).

With the widespread introduction of radio into households in the 1920s, radio became more than another domestic appliance; it became akin to a piece of furniture, or a fine object: "Radio entered most households only after it was domesticated. This meant that it came to resemble furniture instead of a gadget, became easier to operate, and could be enjoyed by more than one person at a time" (Jeffrey 1995, 15).

Covert related how husbands and their sons had to negotiate the terms of reference as to the placement of the radio with the female head of the household:

> In the early twenties articles began to appear in popular magazines describing devious ways of sneaking the radio apparatus into the living room. Females would object. There would be noise, mess, and battery acid on the rug. Therefore, Mother must be lured from home and the radio installed in her absence. In some fashion the unreliable appliance must be persuaded to emit enticing music as she was introduced back onto the scene. She would smile, and the battle would be won (Covert 1992, 304).

By the mid-1920s, advertisers were trying to convince women that the radio was more than just another domestic appliance in the home. The suggestion was made that radio could lessen the drudgery of housework. In North America, Britain, and Australia, broadcasting stations began offering special programs for women, and conceived of women as a distinct audience. Such programs served a two-fold goal: it helped reassure women of the domestic role of this new technology, and also promoted radio's domestic image (Butsch 1998). The Australian magazine *The Listener* printed articles reassuring women that radio was their friend, as programs brought to them household and cookery tips. As Covert writes, the image imparted to women was that

> Wireless was munificent. Women's lives were no longer represented as drudgery and toil, but as normal domesticity. Radio responded to their interests, ensuring that their everyday life was made full by its companionship and variety. Addressed as the fortunate consumers of programs made specially for them, women were "very lucky creatures" (Covert 1992, 65).

Of course, by creating a distinct category of women's programming, radio further reinforced the private and domestic sphere of women. As well, radio advertising convinced the captive audience of women of the need to purchase goods in the marketplace. Programming for men, on the other hand, was encompassing in content, reflecting their public (work, political) lives, and their private (hobby, leisure) lives. [6]

In the late 1920s to mid-1930s, radio programming began to include a distinct category of programming by and for women. Spurred on by notions of scientific household management and the professionalization of home economics, these radio homemakers, typically local women who had a daily show full of help-ful household tips, served the women's community by offering them useful con-tent in a friendly neighborly way. Later, many of these radio homemakers were recruited by advertisers to become spokeswomen for various household prod-ucts. Early North American shows include ones on NBC called "Mrs. Blake's Radio Column," "Sisters of the Skillet," and "Women's Magazine of the Air." Jessie Young, hired in 1926 on radio station KMA in Iowa to host "A Visit With Jesse Young," "became the exemplary radio-homemaker show—a show on which women regularly discussed the details of housekeeping, including cooking and sewing, and also created an easygoing daily radio companion whom listeners could depend on. Jessie didn't lecture; she neighbored. Her voice said, 'Trust me—I'm just like you.' She talked into the microphone as though she were chatting with a friend over a back-yard fence" (Stern & Stern 1991, 84).

Contemporary research, conducted by Peer has looked at whether or not American talk show radio has allowed for a vibrant public sphere for women. Peer concludes that "women's interests, needs, and concerns are not regularly or loudly addressed on talk radio," and that women "do not effectively use talk radio as a medium for conversation and communication and there are far less [sic] women hosting talk radio programs than men" (Peer 2000). Peer blames this fail-ure on the advertiser-driven nature of commercial radio. Douglas also com-ments that American talk radio has created a new gender hybrid—what she dubs the "male hysteric"…"desperately trying to thwart feminism with the reality of having to live with it and accommodate to it" (Douglas 1999). This male pre-rogative of talk radio is epitomized by shock-jock Howard Stern and ultra-con-servative Rush Limbaugh, who are infamous for their diatribes against women and visible minorities—and anyone not white and male.

Women working in public radio broadcasting, e.g., the CBC/Radio Canada in Canada, the BBC in the U.K., and PBS/Pacifica in the U.S., as well as con-tent designed by and for feminist concerns, are research topics in need of vibrant research. Community radio as a forum for feminist voices has been detailed; for instance, Carstensen (2000) has recounted her work at the Feminist International Radio Endeavor (FIRE), a program broadcast in English and Spanish and heard internationally on shortwave Radio for Peace International (RFPI). AMARC (the World Association for Community Radio Broadcasting) and their Women's International Network (WIN) have been working to ensure women's right to communicate through community radio. Their current initiatives include linking community radio broadcasters to one another through the Internet, encouraging women's organizations to own and operate micro power radio, and monitoring the media for women's content. WIN recommends that regular studies of women's participation and representation on the air be conducted, and that monitoring of media regulation and privatization be critically assessed. WIN suggests that women's NGOs be represented at the International Telecommunication Union Technical Committee, which allocated broadcasting spectrum use (AMARC-WIN, April 2000).

GENDER AND TELEVISION

During the 1990s, a flurry of work on women and television was published [7]. What most of these books have in common is a preoccupation with analyzing the multifaceted role of women as audiences in various televisual experiences, with many utilizing an ethnographic approach to contemporary situations. For example, researchers have examined the responses of women to soap operas, talk shows, and sitcoms. This tendency within cultural studies to concentrate on media audiences, and particularly non-elite audiences, has often led to overarching generalizations as to the shaping of subjectivity, audience interpretations, and subcultural resistance to the hegemonic order. More compelling because of their utilization of diverse research materials have been the studies of the early domestication of the television into the North American home.

For instance, Lynn Spigel, in her cultural history of the early integration of the television in the American home, found tell-tale evidence of the history of home spectators in discourses that "spoke of the placement of a chair, or the design of a television set in the room" (Spigel 1992, 187). What she dubbed a "patchwork history" consisted in amassing evidence from popular media accounts that mostly catered to a white middle-class audience, such as representations in popular magazines, advertisements, newspapers, radio, film, and television itself. In particular, Spigel's insistence on treating women's home magazines as valuable historical evidence allowed her to supplement traditional broadcast history (with its reliance on questions of industry, regulation, and technological invention), by highlighting the important role women assumed in the domestic, familial sphere as consumers, producers, and technological negotiators.

Spigel employed a diverse range of historical material to examine how television was represented in the context of the wider social and cultural milieu of the postwar period, such as the entrenchment of women within the domestic arena, the proliferation of the nuclear family sensibility amidst cold-war rhetoric, and the burgeoning spread of single-family homes in the new Levittowns. She culled material from women's magazines, industry trade journals, popular magazines, social scientific studies, the corporate records of the National Broadcasting

Company, advertisements, and television programs.

Spigel argues that preexisting models of gender and generational hierarchy among family members, such as the distinctions between the sexes and between adults and children, and the separate spheres of public versus private, set the tone for television's arrival into the home. As well, the introduction of entertainment machines into the household, including gramophones and the radio, also influenced television's initial reception. Women's home magazines of the time, including *Better Homes and Gardens*, *American Home*, *House Beautiful*, and *Ladies' Home Journal*, were the primary venue for debates on television and the family. The magazines addressed their female audience, not just as passive consumers of television, but also as producers within the household. On the practical side, these magazines advised women on the proper architectural placement for the television set in the domestic space. The television set came to be seen as a valuable household object, becoming an electronic hearth that replaced the fireplace and the piano as the center of family attention.

Television was either greeted as the penultimate in technological advancement and as a "kind of household cement that promised to reassemble the splintered lives of families who had been separated during the war" (ibid. 39); or as a kind of monster that threatened to dominate and wreak havoc on family togetherness. These diverse sentiments were echoed in the advertisements and discourses of the popular magazines of the day. A typical ad by RCA featured the family circle around the television console, while *Ladies' Home Journal* dubbed a new disease which afflicted the young couch potato "telebugeye."

The television industry addressed women as consumers and workers within the domestic economy through advertisements and specialized programming. These discourses addressed to "Mrs. Daytime Home Consumer" included trying to hook the housewife on habitual daytime viewing through genres such as soaps and the segmented variety show featuring cooking and cleaning tips. Women's magazines tried to mediate the dilemma housewives faced between television viewing as a leisure activity and their requisite domestic chores. One absurd solution to this predicament was epitomized by the Western-Holly Company's 1952 design for a combined TV-stove, turning cooking into what Spigel calls a "spectator sport" (ibid. 74).

In the new suburban landscape, television came to be seen as the "window onto the world," and spectatorship became privatized and domesticated. Interior architecture reflected this relationship between the inside and the outside by promoting design elements such as landscape paintings, decorative wallpaper that featured nature or city-scapes, and the picture window-sliding glass doors. Family television show sitcoms mimicked this fixation by depicting domestic spaces in which public exteriors could be glimpsed. As well, through various self-reflexive strategies, such as depicting television characters as real families "who just happened to live their lives on television" (ibid. 158), and through farcical observations on the nature of the medium itself, viewers could be reassured about their relationship with this new electronic medium.

An edited collection, *Private Screenings* (Spigel and Mann 1992), also provides several interesting cases of historical methodology, focusing on the relationship between women, television, and consumer culture. The collection, intended to be part of a larger feminist project of "close analysis and historical contextualization" (ibid. xiii) pays close attention to the analysis of television texts and their historical frameworks. Three recurrent themes are interwoven in the essays. The first is television's appeal to women as consumers, either through its display of various lifestyles and commodities, or through the viewing of television programs. The second theme is memory: how did audiences understand television programs, and what kind of nostalgic function did television programs serve? The negotiation between Hollywood and the television industry is the third theme, whether in early programming where the recycling of Hollywood glitz was common, or through contemporary soap operas which imitate cinematic ploys.

Cecilia Tichi (1991) also examined the television environment as described in the social discourse of popular magazines, cartoons, advertisements, journalism, memoirs, and fiction. Like Spigel, Tichi found that new social rules had to be promulgated with the entrance of television into the familial environment. Television initially disrupted existing norms of etiquette and middle-class behavior. Women had to figure out how best to negotiate the demands of programmed shows versus other commitments to the community and family. The idea of the electronic hearth became pre-eminent; it was a promise of patriotism and domestic security. Tichi argued that the electronic hearth was not politically neutral, since the TV hearth had also been represented as accommodating the community, with friends and neighbors invited to partake of the televised performances and shows—a quasi-theatrical experience in the comfort of one's own living room:

> As the electronic hearth, television is emphatically joined to American history. The discourses of corporate advertisers, media interests, and consensus journalists all evoke in the very term—hearth—the traditionalism of the past. Therefore, television can be claimed as the newest embodiment of values that go deep into the national culture as that culture is historically represented, say, to school students and to an adult public assumed to be middle class in outlook and material means (Tichi 1991, 46).

Tichi further argued that "the hearth embraces male and female realities, serving the women's sphere of family-centered domesticity as well as the masculine forms of patriotic militarism and nationalism" (ibid. 48).

Recent studies on television and women have concentrated on genres targeting women, such as soap operas (Brunsdon 2000) and television talk shows (Shattuc 1998). Women's social issues and how they are talked about in television news and prime-time shows have also been the subject of recent research, e.g., Lisa Cuklanz (1999) on rape and Elizabeth Cole and Andrea Press (1999) on abortion. The introduction of cable television into the American home, and particularly the domestication of erotic or pornographic adult themes by premium channels like HBO, Cinemax, Showtime, and the Playboy Channel, has been explored by Jane Juffer (1998).

Women working in the television industry is a particular blind spot of research; Elayne Rapping comments that "it's another little discussed or acknowledged fact that the lower on the ladder of artistic respectability a cultural form is deemed, the more open it is likely to be to women, racial minorities and gays. Soap operas, for example, are largely written, directed and produced by women...and as those of us who are willing to admit to watching them know, these shows have always been way ahead of prime time—and certainly movies—in addressing issues of gender, race and sexuality from a progressive standpoint" (2000, 20).

CONCLUSION: EXAMINING THE BLIND SPOTS

It is clear that more research on gender and communication technologies needs to be conducted. This includes historical work, using archives (broadcasting, industry, and community resources) and amassing information from popular culture—what Spigel dubs the construction of patchwork histories, utilizing a myriad of texts. It includes assessing the different types of communities created by communication technologies; whether they be communities of receivers, or audience communities as formulated by media institutions and advertisers. We will see later how the entry of the Internet into the domestic environment recalls many of the same concerns that were shown by the domestication of the radio and television, including the reconfiguration of the public and private domains. One of the more compelling blind spots to investigate is the role of labor, and of women working in media industries. The different uses of communication technologies constructed by diverse women, based upon race, class and nationality, can also be explored in more detail. Especially important are the unintended uses made of communication technologies, often made in direct resistance to stated business prerogatives.

Yet another blind spot in research is how women have used the telephone, radio, and perhaps television for purposes of activism. Certainly the telephone and the fax have been used to mobilize women for political work; for instance, in the early 1990s, the Women's Action Committee in New York used telephone trees to quickly spread news (Manegold 1992). Community uses of radio, and not-for-profit public radio, have accommodated feminist concerns, from specific shows to the use of female broadcasters in regular programming. One of the problematic areas in conducting this research is access to the radio programs themselves; it has not been until recently that public broadcasting archives have been deemed to be worth preserving. Fortunately, the Internet is still a public technology, so it is easy for us now to have access to the various communities where women are using the Internet, particularly for disseminating information and collaborating on feminist activism. That will be discussed in the next chapter.

NOTES

1. For instance, see Sandra Harding's work, notably *The "Racial" Economy of Science: Toward a Democratic Future* (Indianapolis: Indiana University Press 1993); Londa Schiebinger's *Nature's Body: Sexual Politics and the Making of Modern Science* (New York: HarperCollins, 1993); Evelyn Fox Keller, *Secrets of Life, Secrets of Death: Essays on Language, Gender, and Science* (New York: Routledge, 1992). A recent study that combines gender and technoscience is *The Gendered Cyborg*, edited by Gill Kirkup, Linda Janes, Kath Woodward, and Fiona Hovenden (New York: Routledge, 2000).

2. A notable exception is Lana Rakow's edited collection, *Women Making Meaning: New Feminist Directions in Communication* (New York: Routledge, 1992) which assesses some of the many themes and issues raised by feminist scholarship in the field of communication since the mid-1980s. The title has a dual purpose: its objective is to show how women have been "active participants in naming the world and making sense of it, even if their contributions and challenges more often than not have been disdained or rendered invisible" (Rakow 1992, vii); and it aims to highlight recent work by feminist scholars whose subjects are the experiences and meanings of other women. Of particular importance is the collection's attention to issues of gender, race, class, and culture, and its inclusion of a multicultural range of female voices. Another more recent contribution is *Gender, Politics and Communication*, edited by Annabelle Sreberny and Liesbet van Zoonen (Hampton Press, 2000). This collection examines the gendered framing of politicians, the coverage of mainstream and movement media for women's movements and how the political concerns of ordinary women's voices are captured in the media.

3. For instance, see: John Brooks, *Telephone: The First Hundred Years* (New York: Harper and Row, 1976); Ithiel de Sola Pool, *The Social Impact of the Telephone* (Cambridge, MA:

The MIT Press, 1977) and *Forecasting the Telephone: A Retrospective Technology* (Cambridge, MA: The MIT Press, 1983); Heather E. Hudson, *When Telephones Reach the Village: The Role of Telecommunications in Rural Development* (Norwood, NJ: Ablex Pub., 1984); Avital Ronell, *The Telephone Book: Technology, Schizophrenia, Electric Speech* (Lincoln: Univ. of Nebraska Press, 1991); Claude S. Fischer, *America Calling: A Social History of the Telephone to 1940* (Berkeley: University of California Press, 1992).

4.   Angela McRobbie has touched upon one of the common conundrums facing feminists who conduct ethnographies on women: intervention. Is it professionally, even morally proper, to influence or interfere with the ideologies and sensibilities of the group being investigated? Can or should the feminist ethnographer attempt to recruit their informants towards their feminist views? How can a feminist academic explore the lives of non-academic and non-feminist women? See Angela McRobbie, "The Politics of Feminist Research: Between Talk, Text and Action," *Feminist Review* n. 12 (October 1992): 3–24.

5.   This trend continues today with the Internet "revenge of the nerds." An early example of this was the hype surrounding Marc Andreeson, one of the creators of Mosaic, the precursor to the Netscape World Wide Web browser. When Netscape went on the stock market in an initial public offering in the summer of 1995, Andreeson was celebrated as a billionaire. Albeit a paper billionaire, he was nonetheless featured on several magazine covers and in photo-spreads with his computer and his girlfriend. Now, more than ever, mainstream media coverage exalts young dot.com millionaires and "entrepreneurs." Internet innovators are featured in the glossy pages of *Vanity Fair* ("The Gods of the Internet"), and even the social and romantic lives of the net.millionaires are fodder for the tabloids, including paeans to their consumeristic fetishes (e.g., Oracle's dot.stud Larry Ellison's $38 million private jet). For a scathing and brilliant critique see Paulina Borsook's *Cyberselfish: A Critical Romp Through the Terribly Libertarian Culture of High Tech* (New York: Public Affairs Books, 2000).

6.   The passive use of the radio by women was also reproduced by the production of content. McKay relates that the use of women's voices in early radio was controversial: "The intimacy of the medium perhaps proved startling to women and men alike, particularly when women performed as announcers" (198). See Anne McKay, "Speaking Up: Amplification and Women's Struggle for Public Expression," pp. 187–206 in *Technology and Women's Voice: Keeping in Touch*, edited by Cheris Kramarae. London: Routledge & Kegan Paul, 1988.

7.   Some of the books include: *Gender Politics and MTV* by Lisa A. Lewis (Philadephia: Temple University Press, 1990); *Women Watching Television* by Andrea L. Press (Philadelphia: University of Pennsylvania Press, 1992); the BFI collection *Women Viewing Violence* (London: British Film Institute, 1992); Ann Gray's *Video Playtime* (New York: Routledge, 1992); *Enterprising Women* by Camille Bacon-Smith (Philadelphia: University of Pennsylvania Press, 1992); Elayne Rapping's *The Movie of the Week* (Minneapolis: University of Minnesota Press, 1992); and *No End to Her: Soap Opera and the Female Subject* by Martha Nochimson (Berkeley: University of California Press, 1993).

3 | WOMEN AS CYBERAGENTS

More of a force for non-violent social change and to serve as a nexus for knitting together global movements to resist injustice, advance human liberation,
and build the beloved community (Coretta Scott King, in an ad for Nortel
Networks' ad campaign, "What do you want the Internet to be?", 2000).

Grrls Need Modems!! (originally from Geekgirl, a cyberfeminist zine
created by Australian Rosie Cross in 1995, and since taken up as an Internet
anthem by many cybergrrls and gals).

This chapter presents several case studies of feminist communication using the
Internet. Here we find examples of women and women's groups using the
Internet for empowerment, edification, and enjoyment. Because the range of feminist communication on the Internet is huge, I have limited this chapter to a discussion of four groups using the Internet to further their communities, particularly
for activism. The first is the use of mailing lists created by feminist academics in
order to create a space to discuss, debate, and promote issues of interest. The
second is the use of the Internet by the global women's movement as a way of
facilitating policy-oriented activism. The third section takes a peek at the cyberfeminism movement, whose members actively use the Internet as a platform for
artistic creation and political critique. The last section looks at how girls are using
the Internet, through e-zines and websites.

It is instructive to go back to the second wave of the women's movement,
when various women's groups were designing and disseminating their own
newspapers, magazines, and books, and starting small press collectives. In North
America, the cheapness and popularity of offset printing led to the creation

about 25 years ago of numerous women's liberation tabloids, including the Washington D.C.–based *off our backs* (still in circulation), *Big Mama Rag, Plexus, The Feminist Voice,* and Donna Allen's *Media Report to Women* (Brownmiller 1999). The notion of what constitutes feminist communication is as relevant today as it was then. Alexa Freeman and Valle Jones wrote that feminist communication must look critically at mainstream media: "we must discard the notion that the media are objective...look at the tools created by corporate media, at the sources of corporate control of the media, and at how we can change that control" (1976, 10). They noted that both the creation and the distribution process is crucial: "feminist communication must incorporate the notion that to communicate is to share, that it is, through whatever medium, a process by which something is exchanged...this principle of mutuality demands both that we have access to the tools of communication and that we actively participate in it" (ibid. 10)

June Arnold staunchly defended feminist small press publishers, urging her readers to reject mainstream publishers: "Madison Avenue publishers, now owned by such as Kinney Rent-a-Car, Gulf and Western and RCA, are really the hard-cover of corporate America, the intellectuals who put the finishing touches on patriarchal politics to make it sell: what we call our finishing press because it is our movement they intend to finish" (1976, 19). In the mid-1970s, North America was home to a plethora of independent women's presses and independent women's bookstores; by the 1980s, these numbers had dwindled, and by the end of the 1990s only a handful of feminist presses remained.

Although many access barriers to the Internet exist for women and women's groups (as will be discussed in the next chapter), the Internet has, without a doubt, been seized by women and women's groups as a relatively inexpensive and fairly flexible tool with which to communicate interpersonally and between women's groups and sympathetic NGO groups.

LISTS: LOW-TECH TOOLS FOR NETWORKING

> Email lists like WMST-L have helped to bring together women around the world, making possible a level of communication and networking unthinkable just a few years ago (Joan Korenman, "Email Forums and Women's Studies," 1999).

One of the earliest and most successful forms of feminist communication on the Internet was, and still is, the creation and use of mailing lists catering to a variety of feminist interests. Joan Korenman's *Gender-Related Electronic Forums* website provides an annotated guide to an eclectic array of over 600 feminist discussion groups (see www.umbc.edu/wmst/forums.html). General topics are organized under 16 sections, including activism, academic topics, health, motherhood, Internet information, women of color lists, and sports/recreation. Within the context of feminist work in the academy, two academic lists, the Women's Studies List (WMST-L) and the Policy Action Research list (PAR-L), illustrate successful models of feminist networking.

WMST-L, the *Women's Studies List*, was initiated in 1991 by Joan Korenman at the University of Maryland-Baltimore County. The aim of the list was to provide a forum for women's studies issues: "a place where people committed to Women's Studies could talk about problems they were encountering in the classroom, exchange syllabi and tips, track down information they need for research, seek help with administering their programs, or learn about relevant conferences, calls for papers, publications, and jobs" (Korenman 1999, 82). One of the first decisions Korenman made was to create WMST-L as an unmoderated list. Later, after user complaints about the heavy volume of daily mail it was generating, Korenman created an edited digest option. To create a cohesive and supportive climate, Korenman has exercised her prerogative as list owner. She decided that the list would only carry discussions on gender-related issues that deal directly with teaching or research, not on general gender-related topics (for more information see www.research.umbc.edu/~korenman/wmst/wmst-l_index.html).

PAR-L, the *Policy Action Research List*, was initiated by the federal Canadian Advisory Council on the Status of Women, and later transferred to the University of New Brunswick when funding for the Council was withdrawn. Its goal is serve as a conduit for developing, conducting, and distributing bilingual feminist research for researchers and activists in Canada. When PAR-L started in 1995, the list consisted of a handful of women, and within one year membership grew to over 300 members. It now consists of approximately 1,000 members, with most in Canada, a significant number in the U.S., and a sprinkling of international subscribers. Action items have included the circulation of letters and petitions to save Studio D, the women's film unit at the National Film Board; strategizing around the closure of Toronto's Women's College Hospital; information about the Multilateral Agreement on Investment (MAI), the Million Women March, and the Global March against Child Labor. Research items have included general queries about feminist research, ethical guidelines for research, and consultation with the federal Status of Women Canada on their Independent Research Fund. Discussions have been wide-ranging: "abortion rights, APEC, backlash journalists, bio-technologies, breast cancer, breastfeeding-friendly workplaces, custody and access, the federal budget, homeless women, lesbians and aging, recovered memory/false memory issues, pay equity, pension inequality, welfare reform, and whether the term web master is sexist (we now use web manager)" (posting from

PAR-L, March 8, 1998). PAR-L also maintains a website, which serves as a resource for policy, action, and research within Canada. It continues to receive support from the Social Science and Humanities Research Council in Canada, and has since partnered with several women's groups, including the Canadian Women's Health Network, Canadian Congress on Learning Opportunities for Women, Canadian Women's Studies Association, and the National Action Committee on the Status of Women (see www.unb.ca/PAR-L).

There are also many electronic support groups for women, and in particular, women's health issues. Galegher, Sproull and Kiesler (1998, 493) characterize electronic support groups as "Internet-based electronic text communication to discuss personal problems or disorders with others who share common circumstances." Breast cancer support lists are extremely popular. Barbara Sharf (1997) analyzed the *Breast Cancer List*, and found a supportive community where members exchanged information, provided social support, and empowered themselves through enhanced decision-making and preparation for new illness-related experiences. Likewise, Carol Devine's Program on Breast Cancer and Environmental Factors Website reports on the links between breast cancer and factors such as diet, alcohol, and exposure to pesticides ("Spreading the Word," 1999).

The issue of women-only lists was controversial in the early days of the Internet, insofar as some men made public their dissatisfaction with the closed nature of the conferences, given the open ethos of the Internet. *Systers-L* was created in the early 1990s by Anita Borg, then at Digital Networks, as a supportive forum for women working in and studying computer science and engineering. *WOW* (Women on the Well) is a forum for women who subscribe to the Well, a San Francisco Bay area network, widely known and popular for its creation of a dynamic virtual community. In order to subscribe to the conference, women had to adhere to a set of rules regarding confidentiality and privacy of WOW conversations, and if there was a question about whether the prospective subscriber was a women, the moderator would telephone the subscriber to make sure the voice was female. Of course, deception could occur, but this was rarely the case.

GLOBALIZATION FROM BELOW:
NETWORKING TRANSNATIONAL WOMEN'S GROUPS

Annabelle Sreberny (1998, 209, 213) suggests that "the decentralized women's movement, using a variety of mediated forms, is a particularly interesting articulation of the new global networking…there are a huge range of women's media and cultural activities on the ground, often ignored by mainstream and malestream commentary, that nevertheless are channels for women's empowerment and thus for social development".

Women's use of communication techniques and technologies for development and grassroots initiatives around the world has been well documented (Gallagher and Quindoza-Santiago 1994; Riano 1994). As Franscesca Miller (1991) has documented, Latin American women have a long and vital history of networking amongst themselves for issues of democracy and social justice, through newsletters, feminist journals, women's movement periodicals, radio programming, and demonstrations, forums, and conferences. But successful networking of this sort depends on a precondition of literacy, and access to the means of production (photocopiers, typewriters, and computers).

Women's coalitions have been created, such as women's presses, information networks, and networks of independent women filmmakers and videographers. Several formal and informal women's communication networks operate regionally and internationally. Regional groups include ISIS International in the Philippines, Fempress in Latin America, Women's Features Service in Asia, FIRE (Feminist International Radio Endeavor), the Tanzania Media Women's Association, and the Association for Progressive Communication. One of the pioneers has been the World Association for Christian Communication Women's Programme, which supports women's use of the media through media monitoring, video production, media literacy training workshops, and gender-based media analysis courses. Recent workshops have been held in Jamaica, South Africa, Cambodia, India, and Nepal. They also provide ongoing critical analyses of the representation of women in media, through conferences such as the 1994 Women Empowering Communication Conference held in Bangkok; and the Global Media Monitoring Project 2000 (www.oneworld.org/wacc/womedia/wp.htm).

The increase and influence of non-governmental organizations (NGOs), especially in the last decade, has been phenomenal. Within the NGO sector, women's NGOs are among the fastest growing segment (Silliman 1999). The range of issues that NGOs have tackled covers a wide spectrum, including development, environmental, health, human rights, legal and scientific areas. In many instances, NGOs have been able to influence government and policy, often thwarting corporate agendas. The use of the Internet has been widely credited as a tool for NGO mobilization (Mathews 1997). [1]

This section will look at a few cases of "globalization from below", where women's groups have used the Internet as an organizational and networking tool to communicate amongst themselves and to influence government and policy. The examples include the use of the Internet for the Fourth World Conference on Women in Beijing in 1995; the use of the Internet to educate the Western world about the Zapatista women in Chiapas, Mexico; and several examples of Internet activism for social justice.

CHAPTER THREE

The Fourth World Conference on Women in Beijing

> La China parece mas cerca desde la pantella de mi computadora. Lamento que
> el acceso a las que no entienden ingles sea mas limitado. Espero que podemosir
> tejiendo estas redes elctronicas tan fuertes que aun con las diferencias de
> idiomas podamos entender a la otra bruja que entra volandopor el correo (China
> seems closer from my computer screen. I am sorry my limited understanding
> of English makes it more difficult to access. I hope that we can build this net-
> work strong enough to overcome these barriers of language so that any "elec-
> tronic witch" can participate). Ana Rivera Lassen, quoted in Gittler (1996).

The Fourth World Conference on Women held in Beijing in 1995 and its concurrent Non-Governmental Forum brought together 30,000 international women whose aim was to provide a second review of the 1985 Nairobi *Forward-Looking Strategies for the Advancement of Women to the Year 2000 (FLS)*, and to provide a platform for action outlining priority areas for concern over the coming five years. The priority areas were poverty, education, training, health, violence, armed conflict, the economy, power and decision-making, institutional mechanisms for the advancement of women, human rights, the media, the environment, and the girl child (Stienstra 1996).

The use of the Internet (through e-mail lists, web conferences, and web-sites) in preparation for the conference, as well as communication to and from the nongovernmental forum during the event, and various post-Beijing follow-up activities, has been a model for many on the role of electronic networking in supporting the international women's movement. Women's NGO groups played a significant role in developing and influencing the Beijing process and policy documents. Their goals were to draft new and alternative language for documents, educate participants about how to influence government, and lobby government delegates.

The Association for Progressive Communication (APC) served as the primary telecommunications provider for NGO groups and United Nations delegates during the preliminary stages of Beijing. The APC itself is part of a larger and interrelated international consortium of peace, environment, and human rights groups that have been using networked technologies since 1990 (Frederick 1992). Networking took place primarily on APC networks, on two mailing lists, *Beijing 95-L* and *Beijing-conf,* and on various World Wide Web sites, including the APC's *WomensNet* and *Virtual Sisterhood* (initiated by Barbara Ann O'Leary). The *Beijing 95-L* listserv focused on pre-conference events and information on travel arrangements, NGO activities, issue-specific causes, and post-conference follow-ups. The *Beijing-conf,* sponsored by the United Nations Development Programme, started originally in preparation for the World Summit on Social Development.

Broad issues of concern related to the conference were discussed, including proposals for agreements and recommendations for implementation and action. Other organizations, such as WEDO (the Women's Environment Development Organization) posted the *Platform for Action* on their site, and Beijing-related documents circulated widely, both electronically and in non-electronic forms, e.g, faxes (Frankson 1996).

The APC's Women's Outreach Program was instrumental in the success of Beijing networking. They provided on-site Internet access at the Beijing conference, policy input to the Platform's Section J theme on new technologies, and continued support after the conference and for Beijing+5 Follow-Up Activities (see Chapter 6). The APC program was initiated in 1993 to facilitate the use of electronic networking for the international women's community. It "promotes women's rights and democratic access to new communication technologies, and supports the empowerment of their organization and networks through the incorporation of computer networking as a tool for coordination, expression and access to information" (Farwell et al. 1999, 103). The program is particularly concerned with redressing the information imbalance between the North and the South by providing training and support activities that can facilitate networking of women through information and communication technologies.

Patti Whaley used the Internet extensively to coordinate the London Amnesty International delegation: "…e-mail was the bread and butter of my daily existence. I looked on the APC conferences and found background documentation, last-minute hotel reservation forms, and gory details about the awful facilities at Hairou" (1996, 226). Whaley's delegation was able to send press releases by e-mail to the Hong Kong office, which then contacted the office in London, which in turn re-posted to the news service on Greennet, the London APC affiliate. Greennet then cross-posted to the Beijing listservs: "Those listservs sent me about 80 messages a day. I had all the major speeches, all the important press releases, and detailed reports of who was stalemating whom. I knew what Bella Abzug was wearing, what days it rained, and how many days it took our list moderator's clothes to dry…I had probably never felt as close to the women's movement as I did then, even though they were literally on the other side of the world from me. Although I work with e-mail daily, I remain astonished and indeed moved at the range and intimacy of the contacts we were able to establish with this technology" (ibid.).

Dineh Davis (1998) monitored the Beijing conference list for ten months before and after the actual conference, and found three manifestations of tolerance: tolerance for a diverse range of opinions; tolerance for continued discussion of procedural variations and violations on the list (e.g., netiquette, list rules, duplicate postings); and minimal flaming (she cites one instance based on ideological and cultural differences). Although widely credited as a successful use of the Internet, there were criticisms based upon unequal access to the technology by developing countries. For instance, by examining electronic networking for the Beijing conference, Ellen Kole (1998) disputes the use of the Internet as a tool for empowerment by women from the South. She found that many women, particularly those in the South, did not benefit through bottom up uses of the Internet as their organizations were not connected to the Internet, due to infrastructure problems (lack of electricity), or because networking was not built into their organization's communication policies. She also found that, although the conference and its participants were international in orientation, the focus of online discussions and conferences tended to be on women in the United States. (Archives are available at www.igc.org/beijing/beijing.html.)

In 1996 the Women's Networking Support Programme conducted a survey of over 700 women's groups and individual women in order to assess the barriers to use and participation of the Internet. The survey had an overall return rate of 21%; of 703 surveys sent out, 147 were returned. The survey revealed that, overall, the Internet was seen as an effective tool for social change work. Positive experiences of using the Internet cited by respondents included: productivity increases, an ability to communicate globally, access to information, and the creation of a culture of information sharing. Negative experiences included time constraints, limited accessibility because of poor technical or social infrastructures, security and privacy concerns, language barriers, and skill deficiencies (APC Women's Network 1997). Although access barriers will be discussed more thoroughly in the following chapters, it is important here to note the disparity between access in the North and the South.

Zapatista Women in Cyberspace

> While indigenous men are often content with their ability to participate in cyberspace, the same is not true for indigenous women. Indigenous websites have, for the most part, failed to include indigenous women. While women have gained stronger representation in indigenous organizations, their opportunities to work with electronic equipment have been circumscribed. Thus, women's voices on the Internet are very limited and access to resources that would facilitate their training continue to be scarce (Guillermo and Becker 1998).

The use of the Internet to coordinate, organize and communicate joint political action has been dubbed "netwar." According to Rand analyst David Ronfeldt, netwar involves a wide range of stakeholders, from the participants "on the ground" to global sympathizers who "use network forms of organization and related doctrines, strategies, and technologies attuned to the information age" (Ronfeldt et al. 1998, 9). However, as Ronfeldt cautions us, netwar does not involve solely information technology—it is also dependent upon old-fashioned communication, such as human couriers and intermediaries. Ronfeldt further distinguishes between netwars and *social* netwars: "social netwars tend to be anti-establishment...[wherein] networks of activist NGOs challenge a government (or rival NGOs) in a public issue area, and the 'war' is mainly over 'information'—who knows what, when, where, and why" (ibid. 21). Social netwars are the sites for public opinion, and for controlling the mainstream media discourse.

The use of the Internet by the EZLN (Ejercito Zapatista de la Liberacion Nacional), more commonly known as the Zapatistas, in Mexico, aided and abetted by international civil society activists and a variety of NGOs, is one famous example of social netwar, which has been well documented (Castells 1997; Cleaver 1998; Froehling 1999; Oppenheimer 1998; Ronfeldt, et al. 1998). Jose Angel Gurria, Mexico's Foreign Minister, was quoted as saying, "The shots lasted ten days...and ever since, the war has been...a war on the Internet" (cited in Mathews, 1997, 54). Rather than reiterating here the complex history of the Zapatista cause, I will very briefly describe the reasons for their insurgency in 1994, the role of Zapatista women in their cause, and the use of the Internet as a forum to inform and educate the Western world about the Zapatista movement, and women's issues in particular.

Chiapas, although rich in natural resources and a major producer of hydroelectricity and coffee, is one of the poorest and most marginalized states in Mexico. It consists of a majority of indigenous peoples whose first language is Mayan. Most of rural Chiapas lacks basic infrastructure, such as water, electricity, education, health care and food. Throughout its history, Chiapas was ignored by the Mexican government. The EZLN was formed to advocate for better and more humane infrastructure; for regaining control of its land from outsiders; and for agricultural reform. The government's economic liberalization policies of the 1980s and 1990s (exacerbated by the North American Free Trade Agreement [NAFTA]) brought about the end of economic subsidies and the regulation of agricultural policies. In particular, then-President Salinas allowed for the privatization of communal land. This led to the 1994 rebellion, when EZLN units took over several towns in Chiapas.

Media reports about the charismatic and enigmatic Subcommandante Marcos uploading his poetic communiquès and other information about the rebellion as a viable means of circumventing the state control of the media captured the public imagination. But it is a mistake to think that the Zapatistas were wired in the jungle. "The EZLN and its communities have had a mediated relationship to the Internet…materials were initially prepared as written communiquès for the mass media and were handed to reporters or to friends to give to reporters. Such material then had to be typed or scanned into electronic format for distribution on the Internet" (Cleaver 1998). The efforts of international activists and NGO groups in the Western world, who were able to translate and post Zapatista messages onto various listservs, newgroups, and websites, increased Western recognition of the plight of the Zapatistas, which led to several national and international plebiscites and intercontinental meetings dealing with the global spread of neoliberalism.

Women in Chiapas also organized and created their own initiatives for rights ("The Revolutionary Laws of Women" and the "Indigenous Women's Petition"). The Revolutionary Laws were a radical departure from the conservative and patriarchal stance that had governed their lives before. Ten statements called for women's right to: participate in the revolutionary struggle, and to share in all the rights and obligations elaborated in the laws and regulations; work and receive a fair salary; participate in the affairs of the community, and to hold positions of authority if freely and democratically elected; receive proper health care, nutrition, and education; choose her own partner and marry freely; and be free from rape and physical mistreatment. The Indigenous Women's Petition called for the government to create childbirth clinics, daycare centers, educational institutions, appropriate transportation, and materials necessary for self-sufficiency (e.g., nutrition, crafts, and markets).

According to Susannah Glusker, women in Chiapas played several important roles. Analyzing information culled from the *Chiapas-list*, she identified six major roles women played: militants, political leaders, celebrities, victim-survivors, academics, and what she dubbed the "invisible stiletto heel brigade." The latter is a group of women working in primarily urban environments. "The 'stiletto heel brigade' struggles from within the system. Their battlefield is office turf, empowerment, employee benefits and facilitating abortion for teenage mothers before it's too late…the network is flexible and energetic…their style is discreet…they work and nurture as individuals from an invisible platform" (1998, 539).

As with the general Zapatista cause, many websites and listservs have been created by Westerners to educate their colleagues about the Zapatistas. Zapatistas in Cyberspace: A Guide to Analysis & Resources, created by Professor Harry Cleaver at the University of Texas, lists a myriad of online conferences, listservs, newgroups, websites, archives, books, and films (www.eco.utexas.edu/faculty/Cleaver/zapsincyber.html). However, only a few websites are devoted specifically to women in Chiapas.

The website Women and the EZLN (flag.blackened.net/revolt/mexico/womenindx.html) provides "texts on the role of women in Chiapas and on their struggle for liberty within that rebellion." A letter from Subcommandante Marcos to the "Insurgentas" (women militias) was posted on the occasion of International Women's Day, March 2000. Also included is Marcos's *12 Women in the 12th Year*—a biographical tribute to women in the movement.

Mujeres Zapatistas (www.actlab.utexas.edu/~geneve/zapwomen) was initiated as a "creative exploratory space" to educate and discuss the role of women in the Zapatista movement. Creator Geneve Gill says she conceived the site when she could not locate other online resources about the Zapatista women. The site is part of a larger effort from the ZapNet Collective, a group of students and activists based at the ACTLAB (Advanced Communications Technology) at the University of Texas, Austin. In her introduction to the site, Gill writes "we have seen that there are people throughout the world who are inspired by the courage and perspicacity of Zapatista women...others are perplexed and challenged by the complexities of the revolutionary articulations of women many times removed from our reality. This is an attempt to bring us together, so that we may learn from you and with each other." She asks for contributions of papers, artwork, photographs, poems, interviews, video, and fieldwork. There is also a Forum on Zapatista Women, which is a threaded message board in English and Spanish. The posts, however, rather than firsthand accounts or recent news, are mainly from students looking for resources or papers on women in Chiapas, often as part of a "hurried research project."

Another initiative is Sisters Across Borders (www.igc.org/ncdm/sisters.html), a network of women in the United States working with women in Chiapas "to break the cycle of poverty and violence imposed on women in both countries by neoliberalism." They have organized Western women's delegations to meet with women and human rights advocates in Chiapas, with the aim of building support for the indigenous women as part of "restructuring the women's movement in the U.S."

One of the inevitable questions that must be raised here is: whose voices are talking about the Zapatista women? Is it Western women's voices only? When Zapatista communiqués are translated from Spanish to English, how pure is the translation? The few websites that are dedicated to women in Chiapas have not been well maintained nor updated. Certainly the physical infrastructure in Chiapas is not conducive to Internet access. Nor is the Internet particularly welcoming to languages other than English.

An uncomfortable relationship can develop when non-indigenous communities appropriate indigenous social movements, or conduct research on these communities (Scheyvens and Leslie 2000). This was pointed out by Nellie Bly in her beautiful film *A Place Called Chiapas*, when the Zapatista cause was taken up in celebratory fashion by many individuals from around the world. Marcos himself was seen as an icon both for young Mexican children who could play with fabricated dolls of him (complete with bandanna and pipe!), and for rebellious webmasters. Ramona (the wizened and toothless spokeswoman for the Zapatistas) has also been made into a doll.

Activism for Social Justice. For years, women in the former Yugoslavia have been using e-mail to further their goals of social solidarity. The Electronic Witches project, initiated in 1994, "has worked with more than one hundred women from thirty organizations throughout former Yugoslavia. These women come from a wide variety of backgrounds, different ethnicities, religions, sexual orientations, education levels, classes, and professions" (Turnipseed 1996). Not only have women in the former Yugoslavia been able to meet their local needs through e-mail, but also they have been able to extend their reach to women in other countries. This section will briefly describe various initiatives that use the Internet for local activism and social justice.

The Network of East-West Women (NEWW) has as their motto: "including the excluded" (see www.neww.org). NEWW is an international communication and resources network "supporting dialogue, informational exchange and activism among those concerned about women's swiftly changing situation in Central and Eastern Union and the former Soviet Union." Founded in 1990 by women from the United States and the former Yugoslavia, it currently links up over 2,000 women's advocates in more than 40 countries working in partnership "to promote tolerance, democracy, non-violence, health, and respect for the institutions of a civil society." NEWW supports the formation of independent women's movements "to increase the capability of women and women's NGOs to intervene effectively on policy regarding women's lives." NEWW principles include the right for women to participate fully in all aspects of public life; the right to reproductive choice; and the right to be free from racial and ethnic hatred, discrimination, violence, and censorship.

NEWW mailing lists are in the English and Russian language. News is culled from independent news sources and independent NGOs. For instance, with respect to Bosnia, there is a Kosova message board, and mailing lists on academic resources, human rights, jobs, and women in war. Their latest initiative, NEWW Online, is an Internet network linking women's NGOs in the Newly Independent States (NIS), Central and Eastern Europe (CEE), Western Europe, and the United States. Its goals are to enable "participants to exchange information, develop technical and institution-building skills, and coordinate research and projects by and about women." E-mail is emphasized as a tool for social change. Participants are provided with stipends to subsidize their use, and in exchange for this, they agree to participate in steering committee meetings, and contribute to and distribute information on the network.

The Women's Environment and Development Organization (WEDO) (www.wedo.org) is an international advocacy network "working to transform society to achieve a healthy and peaceful planet with social, political, economic, and environmental justice for all through the empowerment of women, in all their diversity, and their equal participation with men in decision-making from grassroots to global arenas." WEDO's goals are to influence policy at U.N. conferences, and to monitor the implementation of U.N. conferences. WEDO's 50/50 by 2005 campaign aims to get more women represented in decision making in government, the private sector, the judiciary, media, and academic institutions.

Scarlett Pollock and Jo Sutton of Women'space in Canada have been tireless advocates of women's use of the Internet, through their magazine *Women'space*, and its accompanying website, and also through attempts to influence federal policy in Canada with respect to gender and universal access. Women'space addresses "the enormous resources being spent to get a whole population online without any serious attempt to ensure women and girls are included; the realization that access to the Internet is an equality issue, and that there is a long fight under way just to ensure women have the resources to get online; the need for an inclusive approach, building networks and making the Internet a welcoming and relevant place for a diversity of women where we can work together" (Pollock and Sutton 1998, 42).

Women'space organized the Women's Internet Conference in Ottawa in 1997, which brought together over 175 women across the country interested in using the Internet for women's equality work. Groups included the Ontario Women's Justice Network, which supports social justice issues for women; DAWN Ontario, providing resources for disabled women in the province of Ontario; the National Action Network for Women's Education/Réseau national d'action éducation femmes, an umbrella group consisting of Francophone members and national organizations committed to promoting the French language; the Canadian Women's Health Network, a network of individuals, organizations, and institutions concerned with women's health; Women and Rural Economic Development (WRED), a nonprofit organization dedicated to enhancing the sustainability of rural Ontario communities and promoting economic opportunity through the Internet; the Victoria Women's Sexual Assault Centre; and Pauktuuitk, the Inuit Women's Association, whose aims are to use the Internet to connect women within and between communities to work on issues that affect their daily lives, including daycare, health issues, counselling, family issues, safety, social issues, education for women and youth, Inuit knowledge, justice, gender equality, government participation, and economic development (Pollock and Sutton 1997).

Women'space is also involved in the Women's Internet Campaign, a national campaign to ensure that women, girls, and women's groups have equal access, equal participation, and an equal voice in policy formulation surrounding the Internet. Other examples of feminist activism in Canada include the Candlelight Vigil Across the Internet and the Canadian Women's March Against Poverty, both coordinated by the Canadian Women's Internet Association (CWIA). For the Candlelight Vigil, supporters can download a .gif file (image) of a candle to show support for the anniversary of the December 6th Montreal massacre of fourteen women engineering students in 1989. Clicking on the candle leads viewers to other pages of information about various aspects of violence against women. The World March of Women Against Poverty also includes a "Supporting Wall" website where messages of support can be posted.

CYBERFEMINISM IS NOT...

1. cyberfeminism is not a fragrance
2. cyberfeminism is not a fashion statement
6. cyberfeminism is not boring
19. cyberfeminism is not anti-male
26. cyberfeminism is not separatism
27. cyberfeminism is not a tradition
28. cyberfeminism is not maternalistic
55. cyberfeminisme n'est pas une pipe
83. cyberfeminism is not about boring toys for boring boys
87. cyberfeminism is not caffeine-free
92. cyberfeminism is not lady like
(from *100 Anti-theses*, The Old Boys Network. www.obn.org/cfundef/100antitheses.html)

Cyberfeminism emerged in the early 1990s as a feminist response to technoculture and cyberspace theorizations (Dery 1996). It was at once a stance, an attitude, an emerging theory, a political statement, and an art practice. It is a developing philosophy "which acknowledges, firstly, that there are differences in power between men and women specifically in the digital discourse; and secondly, that cyberfeminists want to change that situation...cyberfeminism is political, it is not an excuse for inaction in the real world, and it is inclusive and respectful of the many cultures which women inhabit" (Hawthorne and Klein 1999, 2).

The Australian feminist art group VNS Matrix articulated one of the first manifestations of cyberfeminism. In their *Cyberfeminist Manifesto for the 21st Century*, they argued for a women-centered technophilia. In a hybrid Haraway-Irigary poetic rant, they proclaimed: "we see art with our kunst, we make art with our kunst—we believe in jouissance madness holiness and poetry—we are the virus of the new world disorder-disrupting the symbolic from within-saboteurs of big daddy main-frame—the clitoris is a direct line to the matrix—VNS Matrix-terminators of the moral code—mercenaries of slime..." (see "VNS Matrix..." 1994).

According to Sadie Plant, cyberfeminism "suggests that there is an inti-mate and possibly subversive element between women and machines—especially the new intelligent machines–which are no longer simply working for man as are women no longer simply working for man" (1997). Seeking to examine the his-tory of women in computing as creators, producers, operators, and consumers, Plant rejects one of the popular conceptions of the relationship of women towards computers as an "anti" stance: technophobic and cautious. Plant likens the his-tory of women and weaving to an intimate comfort with computers. Examining the life of Ada Lovelace, daughter of Lord Byron, who wrote a description of Charles Babbage's Analytical Engine, Plant observed that Lovelace "loved all forms of communication" and that, "at the age of twelve she had entertained hopes of 'writing a book of Flyology illustrated with plates' and told her mother she would 'be able to fly about with all your letters and messages and shall be able to carry them with much more speed than the post or any terrestrial *contrivances*...'" (ibid. 73).

Plant persuasively argues women benefit from the new digital economy. Indeed, Richard Sennett has remarked that "the new capitalism" (where flexi-bility is inherent) "is an often illegible regime of power" (Sennett 1999, 10). Some of the characteristics of work in the networked economy include the increase in the degradation of skilled work, the facilitation of control and power over work-ers by management, and a rise in workplace surveillance. Who wouldn't want to opt out of this? Thus, more women than men have become self-employed entre-preneurs in lieu of climbing the corporate ladder, where the glass ceiling is still a barrier. Women are natural multi-taskers, having learned how to cope with var-ious domestic duties alongside work that has entered the home. Women are also natural networkers, and, according to Plant, have taken to digital technologies with a vengeance, which isn't that surprising—the automation of the work-place, after all, was due to the feminization of the workforce.

Plant draws a parallel between women workers in the textile and cloth-ing industries, many of them recent immigrants working for low wages, with today's electronics assembly jobs, predominantly occupied by young immigrant women who receive low or minimum wages in non-union workplaces. Like their grandmothers before them, who were silent and unacknowledged artists in textile arts and weaving, Plant sees these women as among the first of the digital artists. One of the more hotly contested Internet copyright issues now is not downloading music, but women swapping cross-stitch patterns. "Cross-stitch, as well as the whole needle-works genre of sewing, needlepoint, crocheting, and knitting, has become the passion of a community of techno-savvy Net addicts whose newsgroups and Web sites feed their love of chatting about such activities" (Cryderman 2000). A newly established committee of the International Needlework Retailers Guild will deal with cross-stitch copyright issues.

Many cyberfeminists are also digital artists. Zoe Sofia emphasizes that feminist interest in the digital world "is not based on eroticized relations with the equipment itself but excitement about the possibilities for communication and the exchange of knowledge and eroticism made possible by the emergence of computer *networks*, the prototype of which is that highly feminized technology, the telephone" (1998, 34).

Enter The Old Boys Network (OBN), a group of women digital artists. OBN sponsored the first Cyberfeminist Documenta X, in Kassel, Germany, in 1997. Over thirty women from diverse backgrounds ("from programmers to web mistresses, from artists to theorists, from networkers to geeks and floating in between") and countries (Germany, the Netherlands, the former Yugoslavia, the United States, Australia, Eastern Europe) gathered to stake out a definition — or not — of cyberfeminism. The OBN "is a world wide network of intelligent and sensual super women, having all the possibilities modern life in the western capitalist world can offer. That is: none or too little for them. So they dedicate their hopes, skills, and creativities to Cyberfeminism" (www.obn.org/faq.htm). According to the OBN website, the core group consists of 3–5 women who have assumed responsibility for administrative and organizational tasks, such as the maintenance of the mailing list, website design, organization of real life meetings, and publications.

Faith Wilding believes that the cyberfeminist preoccupation with defining, yet its reluctance to commit to a single definition, has to do with an ambivalence towards past feminist history, theory and practice, and its relevance to the contemporary situation of women immersed in technology. For instance, there is the repudiation of "old style" feminism (i.e., second wave feminism of the 1970s), with its more essentialist, anti-technology, separatist, anti-sex, and often very unhumourous sentiments. Then there is the current wave of cyber-grrlism online, where cyber-utopianism is the creed, stereotypes of women's imagery are ironically adopted and flaunted, yet an actual critical discourse on the cyberworld is absent. Net utopianism is flawed, Wilding says, because "information exchange on the Net does not automatically obliterate hierarchies through free exchange of information across boundaries...the Net is not utopia of nongender, it is not a free space ready for colonization without regard to bodies, sex, age, economics, social class, or race" (1998a). The playful *100 Anti-theses* of cyberfeminism "was an attractive means for engaging conversation, piqueing curiosity, and engaging in language play" but this conceptualist artwork lacked a critical political-economic framework.

Instead, Wilding calls for a definition which embraces activism and a critical sensibility: "Cyberfeminisms can link the historical and philosophical practices of feminism to contemporary feminist projects and networks both on and off the Net, and to the material lives and experiences of women in the New World Order, however differently they are manifested in different countries, among different classes and races." She calls for a critical assessment of the impact of new information and communication technologies on the lives of women, a recognition and a critique of how computer technologies are gendered, and innovative and radical uses and participation with technology ("how can feminist artists and activists create new possibilities and interventions beyond the genderized Net representations of supersexy femme cyborgs or zine cut-and-paste 'Tupperware women'?") (Wilding 1998b).

One of the tasks of cyberfeminism is to reclaim women's technological history. The "Secret History of Women and Technology" chronicled by *Artbyte: The Magazine of Digital Culture* (2000) included Lady Ada Lovelace, who wrote programs for Charles Babbage's Analytical Engine; Grace Murray Hopper, inventor during the Cold War of the first computer compiler; and the OBN of the 1990s.

Cyberfeminist art takes on many forms, but one is a preoccupation with body-centered art: "the vagina and the clitoris have pride of place in much cyberfeminist work...deconstructive projects that address the proliferation of dominant cultural, gender, and sexual codes on the Net will be more effective if they come from a strong, libidinal center and are understood through the filter of women's history and feminist theory" (Wilding 1998b). In a sense, this is a return to feminist performance and body art of the 1970s (Goldberg 1998; Jones 1998). The Web also provides an alternative exhibition space. Women artists have always worked outside the parameters of the traditional gallery spaces. In the 1970s feminists created many alternative art spaces, for example, the Women's Building in Los Angeles. Or artists, such as Martha Rosler, sought different venues for their work—postcards, videos, or public exhibitions such as a 'garage sale' (de Zegher 1998).

One of the many critiques of cyberfeminism is that its interventions are elitist and too intellectually playful, and do not actively engage and challenge the status quo. Does cyberfeminism create alternatives to the current policy milieu? In trying to be as inclusive as possible, is it, indeed, exclusive? Should or can it enter what many call the malestream?

GIRLS, GRRLS, AND MORE GRRLS

Since the mid-1990s, girls' cultural studies have been increasingly popular. Sherrie Inness comments that "far from being a marginal topic, studying girls' culture is essential to understanding how gender works in our society" (1998a, 2). With respect to the media, studies have focused on girls' perception of popular culture, such as magazines (Currie 1999), music, and television shows. Kearney (1998) has looked at girl zine creators, and their anti-consumerist sensibility, and recent attention has also been paid to the creation of girls' games — software created specifically with content designed to attract girls to gaming (Cassell and Jenkins 1998; Vered 1998). It is also necessary to look at adult-produced media for girls (which will be discussed in the next chapter), as well as media created by and for girls themselves. This lets us turn the focus away from consumerist practices of girls, and, as Inness advocates, look at girls as critical agents who are actively involved in the creation of their own culture: "girls do not always swallow the commercialized vision of girlhood whole; instead, they often contest its values and its messages" (1998b, 5).

Here Come the Riot Grrls! The original Riot Grrls were a disparate group of punk feminists who published zines and played in bands, such as Bikini Kill, Heavens to Betsy, and Bratmobile. They reclaimed the word girl with their own feminist twist, asserting a more spunky and can-do attitude, combined with a strong political and activist stance. Rosenberg describes Riot Grrls as "loud", expressing "themselves honestly and straightforwardly" through zines, music, and the spoken word. "Riot Grrl does not shy away from difficult issues and often addressed painful topics such as rape and abuse. Riot Grrl is a call to action, to 'Revolution Girl-Style Now'. At a time in their lives when girls are taught to be silent, Riot Grrl demands that they scream" (1998, 810).

Paper zines, once the mainstay of Riot Grrl communication, have spread to the Internet. Through e-zines, discussion groups, and websites, Riot Grrls "write most often about their days — something small that has upset them or something great that has happened. In that environment, what they create is genuine and accessible. Because the feminism of Riot Grrl is self-determined and grassroots, its greatest power is that it gives girls room to decide for themselves who they are. It provides a viable alternative to the skinny white girls in *Seventeen* and *YM (Young and Modern)* magazines." (ibid. 811).

Nikki Douglas is the editor, publisher, and webmaster of RiotGrrl (www.riotgrrl.com) ("women not scared to be grrls!") and Girl Gamer, for female video and Internet game players, which highlights game reviews, cheats, hardware reviews, and a "Sound Off" forum (www.grrlgamer.com). RiotGrrl features RiotGrrl Interact—a conference system on various topics (sex, rants, books-films-TV, Webology, TeenGrrl, Gen-X), and the Feed the Supermodel game ("as featured in *Wired* Magazine!"). Jennifer Aniston-Pitt was the supermodel, and she could be fed by clicking on several food combinations: Comida Buena (salad, carrots, vitamins, slimfast, crackers) or Comida Mal (hamburger, stromboli, tiramisu, oreo cheesecake, 16 oz. steak).

Confessions of Online Girl Addicts. Girls are using the Internet to communicate amongst themselves, through e-mail, chatrooms, ICQ, and other instant messaging services. Many of them are creating their own personal webpages. Susannah Stern (1999) conducted exploratory research describing adolescent girls' (aged 12 to 17) expression on their webpages. Three main tones distinguished the homepages: spirited (optimistic), sombre (introspective and disillusioned) and self-conscious (mixture of both). Stern concluded that websites help adolescent girls construct their identity by providing an opportunity to openly express their thoughts, opinions, interests and doubts. Stories, poems, essays, and art are the focus of many of these webpages. Many of them are akin to personal diaries, in their honest and funny exposés of their often mundane daily lives:

> 26.09.00 I ran out of milk this morning, after dressing my bowl of honeycomb, so I ran out to the corner store to pick up a carton (as well as two 16-ounce bottles of snapple peach iced tea). I was in my pajamas: beavis and butt-head sweatpants and a stained-with-hairdye t-shirt. I didn't wear a jacket, or a bra, rain fell lightly, cold air blew, and I looked up and the sky seemed more menacing to the northeast, but i felt calm, exhale, smile, don't forget to wear your jacket (from www.maura.com)

Many of the girls' homepages provide a biographical introduction:

> Heylo! I decided to do away with the old bio, again, I'm not very good at writing these stupid things. Hmph. No one reads them anyway :)
>
> I was born on August 23rd, 1980, in Asheville, NC (and that makes me 19 for all you lazy asses). I've lived in Asheville all my life, so, you can bet...I LOVE to travel! Although I've never been out of the US. A few of the larger cities I have been to are: Boston, Chicago, Dayton, Cincinnati, Tampa, Orlando, Denver, Wilmington, Raleigh, Atlanta and DC.
>
> I'm an only child, a "spoiled brat" and I love it. I still live with my parents, in a separate apartment in their basement. So I have some privacy :)

Bryan sez: "Tell them I love you. Cause that's pretty important." (It is). We're officially back together now, I'm so happy! (from www.gerl.org/candy.html)

The creation of e-zines is another popular web activity. *roadrage* features poems, such as this one by Jenyfer:

> **Math in Depth**
> Oh raving man
> Conspirator of Algebra
> Don't slip that equation by me
> No more subtle than a cat stalking prey
> So scary, sending shocks through my mathematically
> challenged soul
> Kill the Negatives!
> Kill the Positives!
> (roadrage.gerl.org/2000/sept)

Geek the Girl zine is created by Athena, "a fourteen year old high school student, but that doesn't tell you much. I'm Filipino-Chinese. My national identity is Filipino. I'm a somewhat willing victim of cultural imperialism because it's the only culture I know. I only speak one language fluently (guess which) but I can get by with the amount of Tagalog/Filipino I know and I can understand a small amount of written Spanish and Italian...I want to go to Berkeley for college, but I intend to live in the Philippines after I graduate. I haven't planned any further than that. I'll probably be crushed if I don't get into Berkeley. I'm not as involved as I would like to be; I'm not an activist, I'm too sheltered and self-conscious for that" (from www.nodeadtress.com/ ezines/geekfirl/memyselfi.html). *Breakfast* is "the zine, not the meal," and features personal notes "with some political stuff thrown in." Articles deal with "Filipino Identity," "Coping with Suicide," "Thoughts on Homophobia," "On Imperialism and Language," and "Racism in my school's administration." Athena also provides a resource list of Asian Girl Zinesters (www.nodeadtrees.com/ezines/geekgirl).

Girls are not just interested in the content of their sites, but in the design and technical aptitude it takes to make an attractive website. For instance, .feisty will host or link to webpages if the following requirements are met:

> You must be female.
> You must have a website.
> No sloppy layouts will be accepted!

"No-no's" include: "broken links, broken picture files, pages that take forever to load, cheesy graphics, frames that don't function properly, pop-up windows, TyPiNg LiKe This" (see gerl.org/feisty).

The Geek Goddess Ring is for "women on the leading edge of the feminization of cyberspace" (www.katygyrl.com/geekgoddess). Sites and self-avowed Geek Goddesses must meet the following criteria for admission to the ring: "Realizes a web*log* is not a web*site*; understands the meaning of quality and content; respects ownership of work and bandwidth; is selective about the number of rings she belongs to; welcomes visitors who either use IE or Netscape; is never located on a server that has pop-up ads; never traps visitors in frames and is easy to navigate; doesn't have ad banners on the front page; values her own opinions and work." Some of the links on the Geek Goddess Ring include:

- Wondergirl: "is like raving to disco!"; features a daily diary. ("Monday, 25 Sept 00: I didn't take a nap today! I've been enjoying my computer too much! My cam is back online! I also took some neat pictures with my faux-pink hair today. I'll post them later.") (www.wondergirl.org)
- Bits of Magic: "Me, computers and the rest" includes digital photos, web-based projects, web design directory, diary (called girl.dot.comp, "the girl of the new bloglennium"). (www.bitsofmagic.com)
- Stop Thief!: "Our Mission: to expose intellectual property theft, including images, layouts, and other copyrighted material, on the world wide web." Run by a group of young female web designers, including Wren (the-junkbox.com), Melissa (littleginsu.net), Andi (screamingmeemies.com), Gillian (soap-bubble.com), and Sarah (janeblush.com), concerned about the lifting of original web designs and content from their webpages. (www.stopthief.org)

E-zines also display a playful and feisty stance. Some of the many include the *Disgruntled Housewife:* "Your Guide to Modern Living & Intersex Relationships." Features include "The Dick List" (highlighting those deserving men); "Ask Queenie" (advice for the lovelorn); "Girls I Like;" "Confessions;" and "Stupid Crap I Bought Last Week" (www.disgruntledhousewife.com).

Other zines include *Maximag* (www.maximag.com); *Geek Girl* (www.geek-girl.com.au); and *Brillo Magazine* (www.virago-net.com/brillo), all of which appropriate and parody women's consumer culture. One of the more striking aspects of women's e-zines and websites is their graphic subversiveness. In a sense, this is a continuation of the guerrilla graphics of the poster and small press designs emanating from second wave feminism (McQuiston 1997), but with a more self-reflexive and pop-cultural twist. As one of the editors of *Bust Magazine* writes, "we can use the power of pop culture to our advantage as well" (Stoller 1999, 272). Targeting the mass media for deconstruction, resistance and ribald humor is one of the elements of third wave feminist media. "Us '90s-type gals like to immerse ourselves in the current pop culture like it's a warm, scented bath. Sure, there's alot of bad stuff out there, but when it comes to magazines, movies, and TV, most of us would rather fight than switch it off. Our favored weapon in this war against warped images of women is the smart chick's version of the kung fu kick: we study it, then slice and dice it, dissecting its context and content to reveal its underlying and often contradictory meanings" (ibid. 266).

This is a tactic that the mainstream media giants who are targeting women and girls as an audience commodity haven't picked up on. To be sure, some of the parodic elements have been appropriated—the use of cartoonish graphics, the branding of "girl power," and mild concessions towards intelligent content—but, for the most part, the feminization of the net, as conceived by the media giants and the new entrepreneurs, remains stuck in the mental swamp of the last century—as we shall see in the next chapter.

NOTES

1. In 1998, the Multilateral Agreement on Investment (MAI), spearheaded by the Organization for Economic Cooperation and Development (OECD), was shelved. Many attribute its defeat to the mobilization of NGOs in using the Internet to post secret documents and coordinate a public awareness campaign. See R.C. Longworth, "Activists on Internet Reshaping Rules for Global Economy," *Chicago Tribune* (July 5, 1999).

4 | COURTING WOMEN@E-COM

"It's the time crunch that's driving women online," explains Bernadette Tracy, President of New York City based NetSmart-Research, a company that surveys Web usage and consults on web marketing strategy. "Women are using the Internet as a timesaving household appliance, just like a washer and dryer." (Kuckinskas 1998).

The rapid commercialization of the Internet and the creation of new services targeted to consumers continue unabated. Some fear that such manic maneuvering by telecommunications firms, new start-ups sped by venture capital, and an emergent digital entrepreneur class will lead to the demise of the Internet as a viable forum for use by civil society (McChesney 2000). Luring Internet surfers online in order to partake of various electronic commerce (e-commerce) services has been a challenge. Guaranteeing the privacy and security of online transactions is a necessary condition; the other is convincing consumers that shopping online is a fun and worthwhile activity. During the early 1990s the percentage of women online hovered between 15–35% (Shade 1994) with most accessing the Internet through institutional affiliations. As the percentage of women online increased, prevailing marketing wisdom dictated that, since women are *supposed* to enjoy shopping more than men, as well as being the primary household shoppers, then designing and attracting services for women could boost the overall success of e-commerce. [1] Commented Lance Rosenzweig, chief executive of Los Angeles–based PeopleSupport: "Women have traditionally been responsible for 80 percent of household purchases. As more purchases take place online, women will continue to take charge in the world as well" (Sandoval May 2000).

Dan Schiller has commented that "there is an outstanding doctoral dissertation to be written detailing the intensive efforts made throughout the last several years to lure women onto the Web" (1999, 183). This chapter will serve as only a modest attempt to document some of these efforts. It will briefly chronicle the emergence and promotion of a commercial Internet environment for women, fuelled by the entry of large corporate interests. While it is not surprising that the women's market would be targeted by commercial interests, most of the services designed for women cater to a middle and upper-middle class and reflect a women's magazine sensibility. The tension between e-commerce applications directed towards women as consumers, and the usage of the Internet as a locus for citizen-oriented activities, as illustrated in the last chapter, provides a useful example of a salient issue the Internet faces as digital capitalism continues unheeded.

DOT.COM

> Along with career and family responsibilities, women have personal needs and interests, from finance to fitness to fashion. They go online to stay on top of current events, get stock tips and of course, find out about the latest fashion trends (Gina Garrubbo 1998).

Electronic commerce is defined as "a modern business methodology that addresses the needs of organizations, merchants, and consumers to cut costs while improving the quality of goods and services and increasing the speed of service delivery." The term also applies to the "use of computer networks to search and retrieve information in support of human and corporate decision making" (Kalakota and Whinston 1996). A burgeoning marketplace of ideas, people, and technologies is developing to support electronic commerce on the Internet.

Electronic commerce brings both promises and problems. It promises to enable companies to shorten procurement cycles through online catalogs, ordering, and payment; expedite JIT (Just in Time) and QR (Quick Response) systems that reduce inventory and facilitate restocking; shrink product development cycles; and accelerate the market delivery of goods and services that can be individually configured for customers. But, this information entrepreneurialism—"the active attempts of organizational groups to take advantage of changes in key social relationships and in information technology to gain market and organizational advantage" (Kling, Ackerman, and Allen 1995)—exists within a system of information capitalism that heightens the potential for consumer privacy breaches.

The Organization for Economic Cooperation and Development (OECD) has taken a lead in promoting e-commerce activities and shaping public policy around key issues such as privacy, security, and interoperability. The 1997 Sacher Report brought together senior executives of major enterprises that use electronic commerce to construct a framework of conditions necessary for the "favourable development of global electronic commerce and the action required by governments at national and international level" (OECD 1997). International events, such as the Ministerial Conference on Electronic Commerce, held in Ottawa, Canada in 1998, have provided a means for dominant countries to identify their "shared vision for global electronic commerce." One of the overarching themes in e-commerce discourse has been facilitation of global economic competitiveness whilst ensuring consumer protection online. The more widespread social implications of e-commerce, such as how citizens become reconfigured as consumers in such an environment, have received little critical attention.

The increase in commercial websites has been characterized by a phenomenal growth rate. In January 1996, a brief perusal of Yahoo's directory of companies brought up 35,373 listings for over 100 categorized companies, from advertising and aerospace, to telecommunications and packaging (Shade 1997a). By September 1999 the number of companies listed in Yahoo was 492,622. September, 1999 figures for domain names registered at InterNIC illustrates the widespread increase in commercial domains; of the 10,910,997 total domains registered worldwide, 6,614,850 were .com domains (see www.domainstats.com).

Schiller's (1999) account of how transnational corporations have seized upon the Internet as the newest market conquest, expedited by global neoliberal policies, provides ample evidence of the need to address the accelerating inequities that are a consequence of digital colonialism. How various transnational corporations (TNCs) and new Internet start-ups have targeted women as a lucrative market to exploit, especially in the e-commerce realm, will be discussed in the next section.

SHOP TILL YOU DROP

> Yahoo cofounder Jerry Yang said he expects the growth in electronic commerce to be fueled by a rise in the number of females on the Internet. He said he expects there will be about 100 million users, with women accounting for half of them. In 1995, the Internet was primarily used in businesses and its 20 million users were 90 percent male, "a geek medium," Yang said (Wong 1998).

It is worthwhile reminding ourselves that, in the earlier days of the Internet, various women and women's groups realized the dynamic potential of networked communications as a way to foster feminist communication across international borders. The use of the Internet by diverse women's NGOs has been well documented (Smith and Balka 1988; Balka 1993; Balka 1997b; Shade 1994) and was discussed in the last chapter. Technological improvements (i.e., from Unix-based to the graphic capabilities of the Web) have made the Internet less a geek and more a popular medium, echoing many of the design attributes of conventional media, such as newspapers and television (Grusin and Bolter 1999).

As more women have come online, with gender statistics almost reaching parity in North America [2], retailers realized that capitalizing on the female market could only boost the success of e-commerce. Industry and cyberpundits began to notice that many women were shopping online during the 1998 Christmas holiday season. In some cases, women accounted for 55% of web shopping (Martin 1998). Active courting of the "elusive female audience" (Time Digital 1999) began in earnest. Reports, including one by the International Corporation, were predicting that "the ratio of men to women online will hit 50-50 in 1999" (ibid.). Katherine Borsecnik, senior vice president for strategic business at America Online, was quoted as saying that "...the early part of the e-commerce revolution was a revolution of rich white males...with 80 to 85 percent of household spending controlled by women, the total retail dollars involved in that change is very high" (Tedeschi 1999). Markets eagerly dissected the female online population: "Using a sophisticated factor analysis, NetSmart discovered that there are no cohesive subgroups of the female online population; women have overlapping, multifaceted responsibilities and interests. They are overwhelmed with multiple demands, their focus based on the current priority rather than a single-minded pursuit" (Garrubbo 1998). However, a few subgroups emerged from NetSmart research. The career women and investor who goes online for convenience and success was dubbed Ms. Biz; while the self-appointed online censor who wears many hats, "from budget gourmet to rainy day fun finder," was dubbed Dr. Mom.

Most observers predict that categories such as apparel, health and beauty, and toys will particularly benefit from a growing presence of women shoppers. Industry media research show that sites offering retail savings and discount coupons for toys and health products (e.g., Coolsavings.com, Toysrus.com, OnHealth.com, FreeShop.com) attract more women, and that "women were less likely to visit such sites as CNET.com and ESPN.com, which are best known for technical content, financial information, sports and news" ("Women Download Savings" 1999; Johnson 2000). Another industry analyst commented that the differences in what men and women prefer online recall the development of niche cable television markets for men and women (Hu 2000).

Predictions are that the online health and beauty products industry, which includes cosmetics, will grow from $509 million in 1999 to $10.4 billion in 2004. A variety of new websites have been created, including beauty.com, beautyjungle.com, beautyscene.com, ibeauty.com, eve.com, and sephora.com (Spiro 2000). In order to attract women to buy cosmetics online, sites are becoming more interactive. Gloss.com features a virtual makeover, where "customers can try on products virtually by choosing a model's image closest to her own" (Olson 1999). The Cosmopolitan Fashion Makeover Deluxe is a CD-Rom that lets women try on clothes online before ordering them from the Macy's Web site. The user creates "My Virtual Model" ("your electronic second self") with her personal measurements. The user can then go into the Dressing Room, where clothes, shoes, and handbags can be dragged and dropped onto the second self (Biersdorfer 1999). In September 1999, Procter & Gamble, one of the world's largest consumer products makers, announced the launch of Reflect.com, an Internet joint venture selling beauty products and services online. Fifty million U.S. dollars have been devoted to the company, "designed to provide beauty solutions to millions of women" (P&G 1999). Style.com was launched by CondeNet, the new media affiliate of the giant fashion publisher Conde Nast, and it prominently features Vogue.com (Sandoval April 2000).

Many media behemoths are creating portal sites for women, where various content, including e-commerce and online discussions, is easily accessible. Not to be left behind on the road ahead, Bill Gates's Microsoft Corporation introduced Women Central, a website for women, in early 1999. The site features content from Women.com, and original material, including Microsoft's *Underwire* on-line magazine. Unilever Corporation, one of the giants in consumer food distribution, is the site's primary sponsor. Unilever's brands include Becel margarine, Lipton Tea, Bertolli olive oil, Breyer's ice cream, Birds Eye frozen vegetables, and Lawry's, Ragu, and Chicken Tonight products. According to Michael Goff, MSN's director of programming, "Microsoft hopes to help women do what they do best on line: stay put. Men tend to surf the Web and jump from site to site, but women are likely to visit fewer sites for specific tasks like joining on-line discussion groups or buying something" (Flynn 1999).

As Internet start-ups and IPOs (initial public offerings) exploded onto Wall Street, it was inevitable that speculators would eye women's sites as a lucrative market. The media paid much attention to the $87.6 million IPO in March 1999 of iVillage, a portal site providing online content focused specifically on women's issues, such as health and fitness, family and parenting, to money management. "iVillage exists as a ganglion of corporate alliances and digital pathways that together form a network of seventeen Web sites, or 'channels,' through which the company seeks to build a community of women on the Internet, and thereby to help them endure divorce, miscarriage, breast cancer, rude children, corporate stress, unfortunate taste in boyfriends, and whatever else a women in the twenty-five-to-fifty-four demographic is likely to experience" (Larson 1999). NBC broadcasting and Intel's CEO were early investors. Specific iVillage interest communities include book clubs, pets, astrology, and a recipe section. Not only does iVillage create its own content, but it aggregates content from partnerships, including Amazon.com, SportsLine USA, AT&T, Infoseek, and Schwab. This is known as the concept of "stickiness" — keeping the eyeballs glued to a particular website and its various links for as long as possible. Sticky devices include e-mail, free personal homepages, horoscopes, chats, and online consultations with experts. In the Parent Soup forum, stickiness is at work when the pregnant woman can enter her due date, see a calendar showing daily bodily changes, chat on the "Pregnancy Circle," click to parenting books at Amazon.com, and shop at iVillage's online stores iBaby and iMaternity. [3]

iVillage's closest competitor is Women.com Networks. In January 1999, a joint venture between women.com and the Hearst Corporation to form Women.com Networks was announced. This venture brought together Hearst's HomeArts.com and its recently acquired Astronet astrology website (which Hearst claims is the top Web site on America Online [AOL]). According to industry communications group Jupiter Communications, HomeArts reaches 2.7% of women at home and work, while Women.com has a 1.2% reach. They are hoping that the merger will create a reach of 11% of women online, and a reach of 6% of the Internet audience (Mittner 1999). Women.com has also gained distribution rights for Hearst's women's magazines, including *Cosmopolitan* and *Good Housekeeping*. Women.com and *Good Housekeeping* announced a partnership to produce an online shopping guide for the 1999 holiday season. The guide, to appear in issues of *Good Housekeeping* and on the Women.com Web site, "will not accept advertising from any site unless it passes the Good Housekeeping Institute's tests for reliability" (Kerwin 1999). A joint venture with E! Entertainment and E!Online for an Oscar-based campaign was announced in March 2000, with an online quiz based on Joan and Melissa Rivers celebrity clothes commentary. The grand prize includes "a trip to Hollywood and a makeover…registrants will be given the opportunity to join Women.com and receive e-mail promotions" (Mara March 2000). Another joint venture with Britain's National Magazine Company (uk.women.com) was announced in June 2000, along with licensing deals in Japan, Latin America, Korea and Malaysia, in order to customize local versions of Women.com's content ("Women.com makes deal" 2000).

One of the more adventurous ventures is Oxygen Media, a combined Internet and cable site, created by Geraldine Laybourne, "considered one of the most influential women in the industry, mostly for her role in building Nickelodeon into a celebrated media brand" (Raik-Allen 1999). Partners include talk-show host Oprah Winfrey, Marcy Carsey, Thomas Werner, and Caryn Mandabach— "the team that created The Cosby Show, Roseanne, and 3rd Rock from the Sun" (Kerwin 1999). Oxygen's websites consist of ThriveOnline (health); GirlsOn (entertainment—films, music, television); befearless (political education and awareness on a variety of issues); ka-ching (money, business, careers); picky (clothes and accessories); BreakUp Girl (love advice and celebrity gossip); handson (women's crafts for sale); Oprah.com, and Oprah Goes Online ("Get Web Smart in a Flash!"). Laybourne

> boasts of her contacts with Madison Avenue heavies like Unilever, and says that the Web's interactivity has a unique ability to build the communities that women seek. She offers Moms Online as a model: "It is very authentic, filled with passion. It's a real community, with 150 volunteers across the country." The Web appeals to women, according to Laybourne, because "women come online to share their experiences. They used to have bountiful sources of community; now they seek that online" (Ledbetter 1998).

In December 1999, French luxury goods tycoon Bernard Arnault bought a $122 million minority stake and a seat on the board in Oxygen (Li 1999), joining investors America Online, ABC, Harpo, Microsoft cofounder Paul Allen, and Vulcan Ventures. In early January 2000, it was announced that Oxygen will supply content for Microsoft's Hotmail through free online newsletters, with content from girlson, Breakup Girl, Thriveonline, Oprah.com, and the "the soon-to-be-launched sports site We Sweat (www.wesweat.com/)" ("Oxygen Media Announces" 2000). One of Oxygen's stranger endeavors is pulse.org, a partnership with the Markle Foundation, which "seeks to learn what women think and believe and to give voice to these findings through the media. We believe that research and the media, used together, can be powerful tools for change." The partnership engages in public opinion research on the information needs and attitudes of women. It is hoped that this audience research can influence cable and web content. In 2000, the focus is on political involvement, participation in the political process, and social and political issues (see Markle Foundation 2000).

NOT A FREE LUNCH

> There are two ways to produce content aimed at women: the 'Wal-Mart' approach, which offers a bit of everything to try to serve everyone, or the 'specialization' approach, which focuses on specific interests (Georgie Raik-Allen 1999).

What sorts of women are being targeted by these commercial websites? Given that advertisers want the most bang from their bucks, it's not at all surprising that middle to upper-middle class women are the targets of sites such as women.com and iVillage. Sandoval reports that women who shop online more than once a week "are between the ages of 45 and 54, earn $75,000 or more, are white, and have children" (2000). In essence, this is the same audience for women's magazines such as *Ladies Home Journal, Redbook, Good Housekeeping*, and *Cosmopolitan*. Most women's fashion magazines (*Vogue, Cosmopolitan*, and *Elle*, for instance) have an online component to complement their print magazine. iVillage's target customers are women between the ages of 25 and 49. The desired audience is an affluent class, including single women, stay-at-home-mothers, working mothers, and professionals. What is interesting to note is which women are *not* courted: certainly not those in the lower socio-economic ranges, and, if network literacy reflects educational attainment, certainly not poorly educated women.

So far, women's content and portal sites are designed for and by the North American market. Hollinger International Inc. is now trying to capture the British women's market, by joining forces with the U.K. drug store giant Boots PLC, by creating handbag.com, a site designed especially for women (Demont 1999). Not only are sites catering to diverse cultural constructions of women, but they are also creating collaborative features to attract women, such as prominent search tools, links to chat rooms, and forums. For instance, Herhifi.com "emphasizes context, selling products by room, like kitchen and home office" (Brier 2000).

Advertisers need to attract affluent audiences, and in the case of commercially sponsored women's websites, market segmentation rules. As Schiller remarks, "current marketing practice portends no carryover into the consumer domain of the principle of equality of representation" (1999, 140). A 1999 Women.com study, done in conjunction with Proctor & Gamble and Harris Interactive, found that "time-pressed women tend to be very focused when seeking information or performing tasks on the Web—very much the in-and-out approach," whereas "men prefer surfing" ("I am Cyber women…" 1999). Advertisers have thus targeted their ads towards women; Toyota Motor Sales U.S.A. Inc. "traded in some of its banner ads to sponsor an area on Women.com…featuring tips on eating right on the road and how to keep kids entertained" (ibid.). Sponsoring content on websites is another way to target women. The "New Mother, New Me" section of Women.com was sponsored by Always feminine hygiene products. Although Women.com designed the content, Proctor & Gamble owned the section. The idea was that such a sponsorship could operate in "stealth mode to catalyse consumer buying decisions…users of feminine sanitary products seldom switch brands, but this special post-pregnancy section was a way to tap into the collective unconscious of thousands of postpartum women and get them thinking in that direction" (Mara May 2000). iVillage's women's auto center is sponsored by Ford Motor Company. iVillage studied their users about cars, determining that their top priorities were safety, maintenance, and leasing. Therefore, the auto section features information from road rage to engineering. Doubleclick Canada, another market research company, is studying how advertisers can attract more women to buy groceries online (Ebden 2000).

Not only are women being targeted online, but advertising in women's magazines for women's sites has also become commonplace. "2 a.m. Restless. Maybe it's work. Maybe it's the uncertainty over Y2K. Maybe it's the future in general. Maybe a new pair of pumps would help you sleep," reads the copy of an ad for Neiman Marcus.com, highlighting the availability of Manolo Blahnik shoes. Attracting the wealthy cybershopper is the goal of many of these ads. LuxuryFinder.com boasts that it features the ".com for those with un.common taste." BestSelections.com is "how the other half clicks" ("shop chic stores in exciting cities — no jet lag, no sore feet!"). An ad for Gazelle.com, a hosiery site, appropriates the imagery of the moon landing, with a panty-hose clad astronaut hoisting the Gazelle.com flag on the moon's surface. The copy proclaims, "One small step for WOMEN, one giant leap for WOMANkind." An ad for Bluefly.com (a discount designer fashion site) shows an illustration of a woman seemingly immersed on her laptop, while in the background a nude male towels off in the doorway of the bathroom, a very large (and erect) cactus adorning the windowsill. The copy: "Satisfaction guaranteed."

The different consumer behavior of men and women online has been a recurrent topic. Ellen Pack, one of the founders of Women.com, said "Men are surfers, whereas women are seekers, going on line to get specific information — advice on acne or how to do a proper sit-up — and not spending hours looking around." A spokesperson for iVillage said "'We have applications and tools that help women get things off their to-do lists'…Accordingly, many women's Web sites are short on news analysis and are instead crammed full of advice" (Teicholz 1998). Another survey looked at the web behavior of men and women in the U.K., France, and Germany. It found that "British men prefer community Web sites where they can communicate with like-minded people…British women, on the other hand, love to shop online; Frenchmen are the ultimate surfers…Frenchwomen, however, have penchant for sites with specific themes such as pagejaunes.fr (France's Yellow Pages); German women like practical services…including Wetteronline.de, a weather service…German men are highly technical, logging on for information about the latest high-tech gadgets" (Koranteng 1999).

Supriya Singh's (1998) research into the gender imbalance of e-commerce in Australia points to the need to conduct further research on how the Internet is being utilized in the domestic environment. Questions she identifies as needing more empirical research include those related to interpersonal relationships within the home; e.g., who exercises and who yields the most power? Is Internet usage in the home related to control of the TV and video remote control? Is Internet usage dependent on the perceptions of men's and women's economic and social valuations? How does usage compare between women in the labor force and those who do not participate in the labor force; and, is domestic access dependent upon socio-economic circumstances and educational attainment, or related to a lack of access to the Internet at the place of work or education?

Whether or not the current crop of women's sites and portals will remain sustainable cannot be known now. iVillage's revenues for 1998 were $15.0 million. Although this reflected a 149% increase from 1997, the net loss for 1998 was $43.7 million, compared with a net loss of $21.3 million in 1997. Women.com's fourth quarter 1998 revenues were $7.2 million, up 157.1 % from the same quarter of the previous year. However, their net loss for the quarter was $13 million, compared with a net loss of $5.1 million in fourth quarter 1997 (see Red Herring Online at www.redherring.com).

THE NEW ENTREPRENEURS

> The Internet, on all levels, is a much more democratic and level playing field. It's more about smarts, drive and passion, than who you are, ethnically and genderwise, said Lisa Crane, CEO of Internet music firm Soundbreak.com (Zeidler 2000).

The Internet industry is composed "primarily of men at the top, with a small but slowly growing percentage of women, especially entrepreneurs" (Hamm-Greenawalt 2000). Many of the companies women own are those they started, such as Petopia, Babystyle, Women.com and iVillage. But, more often than not, women in the Internet industry "work in middle management, particularly in public relations and marketing, with an increasing number of vice presidents. Lots of women work in the trenches" (ibid.). And, despite the dramatic increase in information technology, and the creation of new cybermillionaires, there is still a gender gap in pay ("Tech Industry Nudged…2000). In North America, women were CEOs of approximately 6% of Internet companies that venture capital firms financed in 1999, and they held 45% of top management posts in various start-ups. A broader issue of women working in the Internet industry is how women CEOs and dot.com entrepreneurs are reflected in the media. "The one common complaint of prominent women in Silicon Valley is that, while they are trying to develop and promote exciting new technologies, the media remain obsessively and single-mindedly focused on their looks and their gender" (Brown 1999).

The consensus is that the old rules don't apply. Gayle Crowell, president of e.piphany.net said, "It is not about creating the best products or technology, it is more about brand building and knowing the consumer, and women are great at that" (Kaufman 2000). There is a sense that successful Internet businesses depend on creating brand loyalty and that this is an endeavor which women are good at, perhaps given their more natural aptitudes towards marketing and people skills. "The Internet is a natural place for women," Oxygen Media's Laybourne said, "because they care so much about relationships and staying connected" (Tierney 1998). Perhaps, though, the concentration of women in communication and marketing reflects a reality that women are steered towards these fields, rather than to the technical side of the computer industry.

Women Internet entrepreneurs claim to give women what they desire, such as community and women's culture. They use familiar cultural icons and feminine expectations to ease women and young girls into technology—with girls, it is Barbie's Fashion Designer, and with iVillage, it is the invitation to readers to review films for the "Chick Flicks" section. Or it's about empowerment and a take-charge attitude. Commented Candice Carpenter, CEO of iVillage.com, "We seek to get women all the way from a problem to the solution, like our stop-smoking program. It's not about reading it on the screen. It's about doing it. That's why the Internet is so powerful for women. It's about doing it. It allows women to find her own best life, and to have fun. My mother wasn't driving a Porsche at 60, but I'm going to be" (Lee 2000). This sense of the need for financial independence and attainment of money through digital capitalism is echoed by Kay Koplopvitz, CEO of Working Women Network: "Today's women have it easier — there's liquidity in the market. But venture markets are very high testosterone, very male friendly, where everything is go for the throat. So we will be introducing women entrepreneurs into that world" (ibid.).

Although Geraldine Laybourne, CEO of Oxygen Media, claims that "We don't have a soft gauzy filter over our network. It's not an escapist brand. We are trying to get women out of the ghetto of what's expected of women's content" (Lee 2000), others disagree with her assertions. Comments Macavinta, "A quick scan of the headlines of the most popular women's sites often reveals similar content when it comes to sex, health, and finance —with top features such as 'Seduce Him,' '71 weight-loss tips that really work,' and 'The boss forgot to give you a performance review?'" (1999). Many women feel that they are not part of this mainstream women's audience. Examining recent popular culture designed for women, including television shows, films, books, magazines, and websites, Francine Prose blisteringly comments that "the popular culture being sold to women not only reinforces retro clichés and stereotypical notions of male and female behavior but also is the cultural equivalent of the slightly out-of-date baby formula that, through some regrettable corporate error, gets shipped to third-world countries" (2000, 69). What is passing by our screens, she adds, is "an expensive, seductive, glossy mass invitation to disappear into Cyberpurdah" (ibid.).

Despite the fact that these sites are commercially oriented, there are ways that community has been extended for more altruistic and adventurous means by a new crop of women entrepreneurs. For instance, Evelyn Hannon's journeywomen.com site is a resource for women to exchange information about travel. Says Hannon, "To me, the Internet is such a female thing. Women build communities around themselves and that's basically what the Internet is" (Church 2000). Although she is being courted by several large portals, so far she remains independent, although she is working on several sponsorship deals with bookstores and a resort.

For the most part, women Internet entrepreneurs are creating and developing content within and for the North American market. There are exceptions, such as Qnet (www.qnet.se), a Swedish website targeting women, with content ranging from economics and the law to technology, news, and culture (Johannson 1998). Womenjapan.com targets Japanese women, "giving smart Japanese housewives and women [sic] a chance to participate in the economy" (Cooper 1999). And more Chinese female news reporters than men are using the Internet for newswriting, consulting, and quoting sources (Liming 1998). In Saudi Arabia, it is estimated that two-thirds of Internet users are women. Saudi Arabian businesswomen have found they can bypass the traditional laws and restrictions in that country by using the Internet for buying and selling goods, creating contacts, and working collaboratively (Online in Saudi Arabia 1998). However, such Internet use is most likely conducted in the privacy of one's own home or workplace. Recently, a women-only Internet cafe near Mecca University in Western Saudi Arabia was shut down "for reasons of public morality" (Internet Clampdown in Mecca 2000).

CA$HING IN ON GIRL POWER

Last year, Mattel released Barbie PC, a pink, Barbie-themed computer for girls, in contrast to the boy's blue Hot Wheels PC. While the latter came equipped with a plethora of educational software (including Bodyworks, a program that teaches human anatomy, and a thinking game called Logical Journey of the Zoombinis), the Barbie PC instead came equipped with Barbie Fashion Designer and Detective Barbie (Headlam 2000).

Is Mattel, in a socially conscious move, attempting to lure young girls into computing by creating products that will appeal to them? Or—as seems more likely—will the "pinking" of hardware and the creation of less challenging, girl-identified software and games instead ghettoize girls to their own infantalized cyberculture?

Fostering a more inclusive computer culture that recognizes the different learning styles of girls and boys has been the preoccupation of educators and technologists alike. Recognizing that young women, if not encouraged to enter computing, could become left out of the knowledge-based economy, a commission of the American Association of University Women Educational Foundation released a report delineating recommendations for educating girls in computing skills. The report, focusing on school-based measures, recommended that computers be integrated to work across the curriculum, from math and science topics, to the arts and humanities, and that teachers be educated to become "tech-savvy." Educate students about technology and the future of work, the report said, by emphasizing that computing involves "people skills." Discourage gender bias in educational software and computer games, the report also stated, so that girls can see themselves as designers, rather than mere end users, of computing resources. However, rather than make girls adapt to current computer technology practices, the AAUW emphasized that instead "the computer culture would do well to catch up with the girls...girls' legitimate concerns should focus our attention on changing the software, the way computer science is taught, and the goals we have for using computer technology" (AAUW 2000).

A new crop of Internet entrepreneurs, and others who are making the transition from either broadcast or print media to digital media, have realized that girls are a hot group. Not only are they a demographically large group, but they are also a very large group in terms of purchasing power, being the offspring of the baby boomers, where two-income families and few children are the norm. A bevy of websites have therefore been created to cash in on the tween/teen girls, often through innovative and flashy uses of web-based technology, such as flash, streaming video, and the integration of interactive chat or instant messaging capabilities.

gURL.com was started at the Interactive Telecommunications Program at New York University and is now published by dELiA*s Interactive. This is an example of a site created by young women (in their twenties) for teengirls. gURL is "committed to discussing issues that affect the lives of girls age 13 and up in a non-judgmental, personal way. Through honest writing, visuals and liberal use of humour, we try to give girls a new way of looking at subjects that are crucial to their lives. Our content deals frankly with sexuality, emotions, body image, etc." "Paper Doll Psychology" allows one to dress a figure and receive a pseudopsychological assessment of what the chosen outfit says about the outfitter. "The Boob Files" includes first person essays on breasts, and "Deal With It" tackles all issues regarding sexuality, parents, growing up, and body issues.

Voxxy.com, "not your mother's network" is entering the scene, billing itself as a network where girls can engage in chatrooms, bulletin boards, shopping, and a streaming video series. Voxxy producers have teamed up with Maxine Lapiduss, the former writer and producer of the television shows *Ellen* and *Home Improvement*, and with Hillary Carlip, the editor of *Girl Power: Young People Speak Out*, a collection of interviews of young girls, to produce streaming videos with provocative content. Jennifer Aniston, mega-millionaire star of the television show *Friends* and new wife to heart-throb Brad Pitt, has signed on with Voxxy, where she will develop 13 half-hour programs, interacting with girls on "entertaining as well as empowering" topics. Other shows in the works include a dating game show, where girl panelists ask questions of the boy's mother and sister, and a show called Noise, which explores the "forbidden sanctums of 21-and over Hollywood music clubs" (Anderson 2000). Voxxy plans to extend its brand, creating a cell phone service, a calling card, interactive kiosks in shopping malls, a line of beauty products, and a record label that will sign bands discovered on Voxxy (ibid.).

Cross-licensing and expansion of various product lines are part and parcel of the strategies to hook girls. Synergistic—they hope—ventures include Alloy.com and Penguin Putnam Books for Young Readers partnering to set up a teen book line inspired by poetry and fiction submitted to the site (Sefton 2000). Jim Clark, Netscape's co-founder, is part of a team that has invested $22 million in Kibu.com, a site targeting girls aged 13–18, and featuring free nail polish. Bolt.com offers teen-oriented content and personalized tools such as e-mail, chat, instant messaging, daybooks, and calendars. Content runs the gamut from music, sex and dating, school, and poetry. Bolt.com's revenues come from sponsorships, advertising, and e-commerce. Snowball.com includes ChickClick as one of their networks, which links to 21 sites, including RiotGrrl and GrrlGamer, to Spacegirl, BimBionic (comic strip narratives), and Lawgirl. Chickclick also includes e-commerce capabilities (Chickshop), business and entertainment news (Shewire) and e-mail capabilities (Chickmail).

SmartGirl Internette lets girls rate products, "everything from blue nail polish to J. Crew clothing, from snowboarding to short hair styles." In a nefarious twist, though, SmartGirl sells this information to marketers. SmartGirl's site says: "SmartGirl decided not to show you advertising, so we make money in a different way. SmartGirl does research for companies that make things for girls. We don't tell them what to make—you do! If you see something you like or don't like, now is your chance to speak out about it. Whatever you say, the companies we work with will hear you. Remember, smart girls decide for themselves."

From E-Commerce to E-Commons

As digital capitalism continues unabated, new services and forms of content will be created. Many start-ups and IPOs will undoubtedly fail, while others will reap millions for their investors and creators. The tension between the communitarian aspects of the Internet community and the new vision of unbridled e-commerce is palpable. How can the competing values be reconciled, if at all? The corporatization of the Internet has been presented to us as inevitable. Neoliberal policies, the deregulation of telecommunication services, and a laissez faire, hands-off approach have been presented as necessary for economic competitiveness. While a diverse civil society has been able to utilize the Internet for social justice and democratic well-being, wide access for non-commercial or alternative business use is probably threatened by media behemoths.

Take the case of self-employed entrepreneur Gloria Marinescu. Funded partly by Canada's Department of Human Resources Development under its self-employment assistance program, she registered a Canadian trademark for the name

She-Commerce, and the domain name She-Commerce.net from Network Solutions of Herndon, Virginia (a registrar for Internet addresses). Marinescu discovered that Oxygen had applied for a U.S. trademark for an unhypenated version of She-Commerce, plus similar names, including She-Commerce.org. When she discovered that Oxygen had applied to trademark She Commerce in Canada, she decided to go on the offensive, launching several websites, including www.DearOprah.com, where she criticized Oxygen's conduct. Martinescu admitted, "this was an effort to 'retaliate' because Oxygen was 'going after my She-Commerce' (Partridge 2000). Oxygen later decided not to pursue the Canadian She Commerce trademark, but their lawyers threatened legal action against her because of the "malicious registration" of DearOprah, calling it "a bad faith reaction to a legitimate business dispute" (ibid.).

The $165 billion merger between America Online (AOL) and Time Warner has, for some, been a blatant example of the continuing corporatization of the Internet, with concern being raised about journalistic freedom (Barringer 2000) and a "giant media-Internet dictatorship" (Center for Media Education 2000). Media consolidation and increased power was compared to sexual pleasure, and perhaps conquest, by Ted Turner, Time Warner's Vice Chairman and largest shareholder, who said, "When I cast my vote for 100 million shares, I did it with as much excitement as I felt the first time I made love some 42 years ago" (Hansell 2000).

Arturo Escobar (1999) challenges us to educate ourselves, not just about the political economy of the Internet, but about rethinking a new political ecology of the Internet, which will advance alternative social and political practices. He believes that women, environmentalists, and social movement activists on the Internet will be able to create new cultural practices emphasizing local place-based practices. This is perhaps our best defense against the countervailing tendencies of digital capitalism as we enter the next millennium.

NOTES

1.   *Broadcasting & Cable* reported on a Strategis Group study which reported that "more than half of U.S. adults are now on the Internet, and half of those web surfers are women." The total adult U.S. web-surfing population is estimated to be 106 million. See "More Than Half of U.S. Adults Are Now on the Internet, and Half of Those Web Surfers Are Women." *Broadcasting and Cable* 130(13)(March 27, 2000): 92. Home use of the Internet is also rising, as much as 60% faster than overall usage, including commercial and academic online time. According to a Strategis Group spokesman, "The

Internet is becoming the next big media outlet." See "Women Lead Internet Charge," *United Press International* (March 23, 2000).

2. Measuring the demographics of the Internet has become an ongoing concern for many groups. CommerceNet and the Graphics and Visualization Unit at Georgia Tech have been doing this for years. See CommerceNet Research Center, World Wide Statistics (www.commerce.net/research/stats/wwstats.html) and GVU Ninth WWW User Survey (www.gvu.gatech.edu/user_surveys/survey-1998-04/graphs/general/q29.htm).

3. The fate of iVillage has been the focus of many industry reports. CNET News.com has reported on their first-quarter loss, and their selling of iBaby to online baby-products retailer BabyGear.com. See "iVillage Reports Loss, Loses CFO" (April 27, 2000) and "iVillage in Talks to Sell E-Commerce Baby Unit" (June 2, 2000), URL: www.news.cnet.com.

5 | A GENDERED PERSPECTIVE ON ACCESS TO THE INTERNET

Public policy researchers, educators, and activists have noted the importance of ensuring equitable citizen representation on the Internet. Initiatives to develop Internet platforms in North America, including those formulated by the United States Advisory Council on the National Information Infrastructure (NII) and the Information Highway Advisory Council (IHAC) in Canada, and current discourse on the digital divide illustrate that we are at a crucial juncture in examining and elaborating policies of universal access and service. It is extremely important that citizens become cognizant of evolving policy issues surrounding the Internet and have the opportunity to become involved in policy debates.

Issues of inclusiveness to the Internet are crucial because they are issues of fundamental democracy. Universal access to communication and information services must be recognized as an essential human right for maintaining basic democratic values. The premise that communication and public access is an economically justifiable public good, where the most benefit to all accrues when all citizens and interconnections are made, needs to be vigorously maintained. This is especially important as the debates continue over the encompassing factors of universality in a social environment characterized by rapid technological convergence, competition, commercialization and deregulation. As Benjamin Barber has asked, "what are the market incentives to protect public interests" amidst an increasingly "monopolistic infotainment sector" (Barber 1995, 86).

Although those formulating public policy have been slow to focus on gender as a variable, debate and discussion have emerged on the gendered implications of access. This chapter aims to further this important area of research and debate, particularly within the North American context. The section "Gender Troubles" focuses on the myriad issues of access, including economic and physical access, as well as workplace issues surrounding digital technologies. The many issues surrounding substance and content on the Internet will also be covered. These issues run the gamut from user-centered design practices to the controversial issue of online content and ethics, particularly that of online sexual harassment and pornography.

The third section of this chapter, "Proposed Solutions," considers some solutions to amending these gender troubles. It looks at physical access—community access points and domestic access, funding and support mechanisms—and argues for the sustenance of community networks. The creation of acceptable use policies and the development of user-centered design are also examined briefly, as are public policy statements on adding gender equity to the information infrastructure.

"Looking into the Future," the fourth section, evaluates the proposed solutions, maintaining that as Internet policies and technologies grow, humanistic values should accompany their inception, design, development, and diffusion. Such humanistic goals include ensuring equitable access, a diversity of viewpoints, heterogeneity of content and information, and community and democratic enhancement.

The chapter concludes by reasserting the importance of a gender perspective and how high the stakes are. Issues surrounding democratic participation and edifying expression must be upheld. The challenge that we face now is in ensuring that in our current rush to privatize, deregulate and commodify the information infrastructure, we face the challenge of sustaining and widening the social infrastructure to encompass a diverse citizenry.

GENDER TROUBLES

Access. Women's relationship to the computer can be characterized, simply, as problematic, as shown by an extensive literature on the gendering of computer technologies. Research has concentrated on the following issues: what uses have computers been designed for? what language has grown up around their development? and what social practices have been brought about by computers, in workplaces, academia, and in popular culture? [1]

Dale Spender echoes the remarks of many feminists when she writes "The glass ceiling may be preventing women from getting into the top levels of general management, but it is also preventing them from getting into cyberspace in appreciable numbers. Yet this is where the new communities are being formed; this is where the new human values are being forged" (Spender 1995, xxiv).

Various demographic studies exploring gender representation on the Internet have been released since the mid-1990s, with most of them conducted for commercial and market research. Mid- to late-1990s statistics fluctuated, but all of the studies indicated that women's access, although not equal to men's, was slowly increasing. Despite these lower figures for female representation, there was an optimism that usage amongst women would increase. For instance, the Rand Corporation, in a study examining universal access to e-mail, determined that "on the whole, our analysis of sex differences in access to information and communications technology provides evidence that the gender gap among adults has decreased; and we concur with the *Times Mirror* conclusion that it could disappear entirely in the next generation" (Anderson et al. 1995). [2]

As discussed in the previous chapter, various commercial interests have been developing content aimed at women to encourage women to use the Internet to shop. An August 2000 study conducted by Media Matrix, an industry group that monitors computer use, revealed that in the United States more women than men are online. The report stated that, "while the universal online population increased by 22.4 percent [in 1999], female users increased by almost 35 percent" (Laucius 2000). The category growing most was girls aged 12–17, estimated up more than 126%, with girls aged 2–11 up 53%, and women aged 55 and over up by 110%. A study from the British Department of Trade and Industry found that, in a 3-month period in 2000, 40% of new Internet users were women. Predictions are that women aged 25–40 will account for six out of ten Internet users by the end of 2000. The ten most recognized websites cited in the study included Amazon.com, Egg.co.uk, Handbag.com, Letsbuyit.com, and Readytoshop.com (www.theregister.co.uk/).

Anthony Wilhelm suggests that virtual political public spheres need to take into account the antecedent resources, "the skills and capacities that one brings to the table to achieve certain functions" (2000, 35). These antecedent resources encompass a range of socioeconomic aspects, including occupation, education, and income. When looking at which groups have disproportionate access to these resources, we find race, ethnicity, and gender matter. So, despite the very recent optimistic figures on gender access to the Internet promulgated by industry and commercial groups, "digital divide" statistics provide a more troubling perspective.

Figures released by the U.S. National Telecommunications Information Administration (NTIA) in 1999 show a persistent digital divide. Urban households with incomes over $75,000 per year are twenty times more likely to have access to the Internet than rural households at the lowest income level. Households with higher education levels are far more likely to have Internet and computers in their house than those with the lowest education levels. Single-family households, particularly those headed by a female, are less likely to own a computer (31.7%), compared to married couples with children (61.8%). The same is true for Internet access: for single-family households led by a female computer ownership is 15%, while for dual-parent households it is 39.3%. Female households with children under the age of 18 have the lowest access to the Internet; domestic access is a mere 10.6%, and access outside the home is 14.7%. When compared to male households with children under the age of 18, these figures are not that significant (13.5% and 15.2%, respectively), but when compared to married couples with children under the age of 18, there is a significant difference in domestic access (27.7%). The NTIA report also noted that "the digital divide has turned into a 'racial ravine' when one looks at access among households of different races and ethnic origins", with the gap between White and Black household Internet access in the home increasing 37.7% in one year alone (U.S. Department of Commerce 1999).

In Canada, a study conducted by the Public Interest Advisory Centre and funded by Industry Canada and Human Resources Development Canada (Reddick 2000) looked at the characteristics of those Canadians who are not connected to the Internet. Men's usage of the Internet was slightly higher (56%) than women's (45%), but females are overrepresented in the non-user population. Because access varies based on social class, generation, and value factors (perceived need and interest), it can be assumed that women have lower rates of education, income, and perceived need. Indeed, a Statistics Canada study on women in Canada reveals that, although females account for 50.4% of the Canadian population, this equality in numbers does not correspond to socioeconomic equality for women. The female to male earnings ratio was 73% in 1995, and the earnings of women were significantly lower than men in all occupational categories and at all levels of educational attainment. Furthermore, unpaid work, part-time and nonstandard work, a lack of upward occupational mobility, and a tendency to concentrate in low-wage clerical, sales, and service sectors creates an additional burden on women and can help explain why women access and participate in information technology in lower numbers than men (Statistics Canada 1995).

Work Issues. In evaluating women's relationship to the Internet, one of the more important factors is women's relationship to paid work. Claims by management gurus and technological pundits that the information infrastructure will create jobs and accelerate the trend towards a knowledge economy replete with lifelong learning opportunities need to be investigated. Particular issues to examine include:

- What jobs are being deskilled by the introduction of computerization and networked communications? For instance, if libraries are being wired, how is this affecting a workforce predominately staffed by women? (Harris 2000).
- Are women using networked technologies in entrepreneurial ways? Are women significantly involved in the commercial end of networking (i.e., as consultants, owners of businesses, content creators)?
- What are the obstacles (e.g,. educational barriers, lack of affordable childcare) facing women who seek to enter and thrive in the high tech fields?

In particular, the impact of telework on women needs to be explored. Oldfield (1991) writes that the triple workload of women teleworkers (paid work, housework, and childcare) creates added stress, dependance on a spouses' wages, an increased risk of poverty, and isolation. Oldfield suggests that narrow-scope public policy strategies for improving the lives of women teleworkers include ensuring that: homeworkers are guaranteed employee status by union membership; homeworkers are included in a broadening of legislative protection such as through unemployment insurance benefits; and that real-life and virtual networking organizations are created, whose goal would be to reduce home isolation and disseminate public awareness information for teleworkers.

So far, public policy has not addressed the broader social issues of telework (Borowy and Johnson 1995). For instance, telework has been promoted as a way for women to work at home while performing childcare responsibilities. This promotional stance does not recognize the added stress that juggling these dual roles can create for the employer as well as for the teleworker and her family. The creation of flexible childcare arrangements for the teleworker who may need erratic or less than full-time care needs to be addressed. This attention seems unlikely, however, given the dismal record in North America of investing in quality childcare at the federal level, and the trend towards eliminating support (at state or provincial levels) in favor of privatized or ad hoc arrangements.

Oldfield refers to the "not-so-hidden agenda behind telework: mandatory self-employment" (Oldfield 1995, 16). Indeed, many telework because this is their only option in a jobless economy littered with contract and part-time work. Telework accords with turn-of-the-millennium ideal of virtual corporations, staffed by virtual employees who can expedite models of electronic commerce. Women who telework could find that the gains they have made in the workplace have been eroded because their virtual presence isn't as important or impacting as their real presence in the office. The obstacles to cracking the virtual glass ceiling are likely to be overwhelming.

However, some women find that the benefits of telework, such as flexible work and social arrangements, outweigh the disadvantages. As both governments and corporations downsize, work and workplace are being reconceptualized by technological advances. Given the trend towards the redomestication of many forms of work, from low-level clerical entry jobs, to professional symbolic analyst work, many women find that this flattening of hierarchies is advantageous for them.

For good or for ill, women are part of the new digital environment. It is important to measure how women are actually using computer networks and online services. What functions and services are they most likely to use? What kinds of content do they desire? What impedes their use—economic factors, technical factors, or a lack of interest? Resnick's (1995) survey of women online revealed that e-mail is the Internet feature women use most—typically for connecting with co-workers and friends in distant locales. Access to special interest bulletin boards with topics like feminist issues, socially responsible investing, small business matters, parenting, and legal issues ranked a close second. Surprisingly for the mavens of video-on-demand and the proponents of interactive cybermalls, shopping was ranked as the least-used online service; two-thirds of the women surveyed had never purchased a single item of online merchandise.

Five years later, however, the Media Matrix survey mentioned above revealed a surprising dearth of communication for feminist goals. Canadian spokeswoman Sherry Barmania said women are "more likely to be looking for answers to questions about health, social activities, travel and financial planning" (Laucius 2000, A2). The Media Matrix survey showed that women in the 25–34 age category tend to visit Pampers.com, various children's and maternitywear retailers, and the birthday planning site, Birthdayexpress.com. Women in the 35–44 age category visit, as their top website, alka-seltzer.com; women 45–54, onehanesplace.com (shopping for Hanes hosiery); while women 55 and over visit the website of the American Association of Retired Persons. College-aged women (18–24) visit bigwords.com (a seller of textbooks, CDs, and clothing), while their younger counterparts (aged 12–17) cite cosmogirl.com as their favorite website ("Graphic Evidence" 2000). These choices for top websites visited coincide incredibly with the commercial sites mentioned. Research methods need to be questioned. However, the Pew Internet & American Life Project described women's favorite Web activities as looking for health information, checking out job information, playing games online, and hunting for religious or spiritual information (Pew Internet 2000). This consideration of content leads us to the second aspect of gender troubles.

Issues of Substance and Content. Access for women encompasses issues of content. North American cases have highlighted how gender has affected the development and cognizance of emerging social mores surrounding the creation and sustenance of the Internet community. These cases have included issues surrounding interpersonal conduct (or netiquette), sexual harassment, privacy, anonymity, identity, and free speech and pornography. The clash between evolving net norms and new strategies to resolve, police, or regulate behavior and content on the Internet by policymakers, governments, and communities has created tensions and acrimony over the future of the Internet (Shade 1997c).

In order to encourage more women to get and stay online, the development of online gender information services must be promoted. The kinds of information and resources that can and should be provided for women should be identified, with a plurality of viewpoints respecting the tenets of free speech represented. This can include information that resides in the federal, provincial, municipal, and community level.

Identification of basic content services for women and women's groups should be conducted (e.g., community-wide women's health clinics, birth clinics, abortion clinics; counseling; woman-oriented small business services; daycare and childcare providers and centers, etc.). Also, what sorts of information should be deemed essential for education, public health, or public safety (e.g., adult education centers; women's centers; community health clinics; local Planned Parenthoods; women's shelters; information for battered women, etc.)?

Issues of sexual harassment and pornographic content online also affect access for women. Bell and de la Rue characterize online harassment as open hostility and harassment (sexual or not) directed towards women by men: "[The] open hostility towards women seems to be analogous to the experience of women entering traditionally male-dominated professions and trades" (1995). Online harassment more often occurs via private e-mail and in IRC (Internet Relay Chat) areas, and the nature of computer-mediated-communication (its slant towards anonymity, lack of personal contact, etc.) tends to encourage flamboyant, outrageous or nasty behavior (Brail 1996).

However, as Jodi Dean has pointed out, it is necessary to avoid the stance of victim feminism. "Scary misconceptions of the Net...[can] prevent women from finding and using the already vast resources of the Net. Regulation is manifested on the Internet as the fear that networked spaces are necessarily hostile to women. It appears as the border that keeps out women, as well as racial and economic minorities" (1999, 1071). Wendy Kaminer (1998) also cautions against public policies on Internet content that are given new feminist spins; she cites the case of the Loudon County, Virginia, library board voting, in 1997, to equip all publicly accessible computers with filtering software so that children and adults could not view pornography or obscene material. The policy, later overthrown in a court, was titled "Policy on Internet Sexual Harassment."

Issues surrounding privacy and harassment related to the Internet are another topical concern. For instance, the furor over of a World Wide Web site titled "Babes on the Web" raised the issue of whether or not linking to the personal and professional Web pages of women, which were then rated on the basis of their personal photographs, constituted harassment, violations of privacy, or merely annoyance (Shade 1996).

The debates about pornography on the Internet, fuelled by the Communications Decency Act, [3] tend to reflect the same sentiments as the unabated and impassioned pornography debates between feminists, with antipornography feminists contending that pornography is sex discrimination on one side; and "free speech" feminists, who distrust censorship, on the other side. Catharine MacKinnon, a forceful voice in the former camp, has continued her crusade against pornography on the Internet: "Pornography is a huge amount of activity on the Internet, which aspires to be a universal network to unite the world. When men make new communities, they bring their pornography with them. More than that, they bond through it. Computer networks are only metaphors for society, they track it and happen within it" (MacKinnon 1995).

Gender troubles may arise even when there are no malevolent intentions.

Although the predominance of masculine values may be largely unconscious, these values are widely held to be the universal human norm. The mythology of computer culture, as Karen Coyle (1996) comments, subscribes to a powerful heroic and machismo mentality.

J.C. Herz (female) wrote, regarding the presence of women in cyberspace:

> Forget the media ballyhoo about electronic town halls and virtual parlors; the net is more saloon than salon. Not too many women in these here parts, scant discussion of philosophy and impressionist paintings, and no tea sandwiches. Rather, much of the Net exudes a ballistic ambience seldom found outside post-apocalyptic splatterpunk video games. Someone should nail up a sign: 'Now entering the Net. Welcome to Boyland...' (Herz 1995, 52–3).

Herz's humorous embellishment on the typical Internet denizen is clearly overstated but it does raise an important issue. Joan Greenbaum (1990) and Bodker and Greenbaum (1993) have analyzed the gender perspectives that underlie the systems development process, which typically ascribes male values (objectivity, impersonality, rationality, power) versus female values (subjectivity, personal feelings, emotions, love) to good system design.

Addressing the problem of getting more women onto the Internet includes tackling not only their relationship and accessibility towards the technical (hardware and software) accoutrements, but also to the social infrastructure and relations, including the complex issues surrounding content, offensive speech, and an often inhospitable environment. The next section will consider some proposed solutions to these gender troubles.

PROPOSED SOLUTIONS

Access to the Internet is multifaceted, and encompasses physical, technical, economic, and social factors. Public policy examinations on access mainly consider the technical barriers towards access, for instance, the hardware and software to support communication, resource discovery tools and issues surrounding interoperability. The myriad factors that comprise the social infrastructure need to be considered as a holistic component affecting the access triumvirate of equity, affordability, and ubiquity. These include an ongoing examination of the many facets of network literacy; and of the diverse social variables affecting geographic, linguistic, income, gender, and class-based barriers (Clement and Shade, 2000).

The creation of public access network sites in community centers and public libraries is a necessary requisite to meet universal service goals

(Skrzeszewski and Cubberley March 1995a). However, true access and ubiquity will not be attained until networked technology is more easily and economically brought into the home. Such domestic ubiquity will significantly increase women's access to the information infrastructure.

There are many reasons for this. Despite the rhetorical push in current public policy towards creating public access points at public libraries, schools, hospitals, post offices, and community centers, these access sites are not readily available now. Even if these sites were ubiquitous, many women (particularly those with small children, the elderly, and the disabled) might find it difficult to get out of the house. Because the asynchronous nature of networked communications makes for cost- and time-efficient communication, the argument can be made that networked computers will be at least as efficient to use as other widely used household communications tools, such as the telephone and the videoplayer.

How can these services be supported so as to encompass the goal of universality? What sorts of information safety nets can be designed and established so that all citizens, regardless of their ability to pay, can partake of services? A variety of funding options is available for support mechanisms, but many questions remain to be answered. For instance, should telecommunication carriers be required to contribute to a universal access fund? Should tax credits be made available for telecommunication carriers that contribute to a universal access fund? Should the telecommunications and computer industries be encouraged to develop a standard information appliance (akin to a device attached to the television, or a Minitel/videotex box) which would allow low-income users to access the Internet? And how can nonprofit public spaces be created and sustained in an era of increasing commercialization? Sustenance of a public sphere broadens the possibilities for a range of citizens to participate in the benefits of the information infrastructure. This can potentially ameliorate the distinction between the information have's and have-not's, and also extend and enhance democratic practices.

This is where community-based computer networks (or "free-nets") can play a vital role (Schuler 1996). An excellent example of how women can get involved is illustrated by the efforts of Pat Nelson, Co-Chair of the Edmonton Free-Net Content Committee, to create a support system for women. In addition to rallying local community organizations to develop content of interest to women, she has also conducted free training sessions specifically for women. The first workshop had over 200 participants. It not only enlisted a plethora of new subscribers to the free-net, but also generated a host of content providers.

Nelson said of the women workshop attendees, "Once you get them to water, they're eager to drink" [4].

Another emerging field of study is Community Informatics (CI). According to Michael Gurstein (2000, 3), "CI studies how ICT (information and communication technologies) can help achieve a community's social, economic, political, or cultural goals." CI links economic and social development with community efforts at the local level, for electronic commerce, electronic democracy and civic participation, advocacy, cultural enhancement, and other uses.

But it is also important that content cannot only be tasted but created; a one-way broadcasting model of communication needs to be converted to interactive and dynamic resources, where users can be creators, collaborators, and contributors to the content.

Developing User-Centered Design: Several multimedia artists have been experimenting with feminist design aesthetics. Christine Tamblyn argues that "because computers have evolved as tools built by men for men to be used in warfare, the current interfaces tend to have a violent, aggressive character. They are hierarchical, mirroring the militaristic male pyramid with its rigid chain of command" (Tamblyn 1994). For Tamblyn, "interfaces designed to be operated by women ought to be multi-sensory, personal, affective, and dynamic." She suggests that the interface feature "a female persona in cyberspace who serves as a guide to the system. The navigation buttons on each screen appear inside an image of this persona, and her voice gives instructions about how to proceed" (ibid.).

Likewise, Joan Truckenbrod argues that a feminist design aesthetic "involves the development of kinaesthetic, holistic, accessible, interactive computing technologies that advance expressive as well as instrumental values. Such an aesthetic would challenge the existing single-minded, malestream commitment to what Mumford calls 'authoritarian technics,' a system created for power and control" (Truckenbrod 1994).

However, we must be wary of essentialist stances which posit that technologies must be based on universal feminine attributes and values. Forging beyond masculinity and femininity to create technologies that subscribe to new social values and needs is an imperative task, but these new social values and needs can and should be multifaceted. A range, as well as an openness, of aesthetics is important if we wish to maintain a sense of inclusivity. [5]

The gendering of computer systems design and feminist approaches towards computer systems design have been the focus of a body of research and theory. The late Margaret Benston was one of the first to consider how a feminist approach would differ from conventional design processes; and for her this "is the recognition of the need to involve the people who will be using a technology in its design, with the aim of incorporating more humane values in such systems" and also of the need to identify and provide an analysis "of the role of experts and authorities in scientific and technical work that comes out of a feminist analysis of masculinity and control in science" (Benston 1989, 207).

Suchman and Jordan (1989) have argued that incorporation into technology of the everyday work practices of the users of that technology is imperative in order to design appropriate technologies. In particular, the design of technologies that are sensitive to women's knowledge and concerns and work practices is necessary. For instance, early office automation systems were designed without the input of the workers who operated the systems on a daily basis—typically women. Often computer systems and chairs were set in fixed positions, thus not allowing for the physical flexibility needed to accomplish a multi-faceted job.

Participatory design (PD) practices, where the users of the technology initiate active participation in the systems design of the computer systems (Schuler and Namioka 1993), have not explicitly addressed gender as a factor. However, Ellen Balka's research into designing computer networks for feminist non-profit organizations has introduced an analysis of gender as a factor in PD projects (Balka and Doucette 1994; Balka 1997a). Balka suggests that notions of PD must be expanded from considerations of business norms to encompass a wider range of organizational settings, such as organizations that are unstable or poorly capitalized, and that include a wide range of learners. As well, "participatory design within women's organizations is likely to fail unless the gendered nature of expertise is recognized, and specific procedures are introduced into the PD process that reflect the difference in women's and men's learning styles, as well as the gendered nature of expertise" (Balka 1997a).

What would women-centered or women-friendly system design look like? Certainly it is difficult to generalize here (and such generalizations also run the risk of positing essentialist arguments), but several design issues (both technical and social) could be considered to be women-centered or friendly. These include the development of infrastructure applications that support strong security and privacy.

Many women and women's groups are very concerned about maintaining the privacy and integrity of their networked communications. In particular, many women's groups (including shelters for battered women, hostels for homeless women, telephone crisis lines and rape crisis centers) deal with issues of a very sensitive nature, i.e., battering, harassment, sexual assault, incest, and child abuse. The need for security on computer networks is of paramount concern for the women's sector. The development of infrastructure and applications for supporting strong security, and the use of encryption and digital signature technologies to ensure the privacy and authenticity of communication, is one technical method to ensure privacy (Dam and Lin 1996).

The Internet must be easily accessible to the broad population, including people with disabilities, and multiple access methods are necessary to account for differences in human capacities. Ben Shneiderman recommends that more attention be paid to "identifying appropriate services, designing a consistent user interface, and developing a clearer model of the diverse user communities" (Shneiderman 1995, 162). Human-computer interaction research and usability engineering must become sensitive to the broad range of user diversity, which "involves accommodating users with different skills, knowledge, age, gender, disabilities, disabling conditions…literacy, culture, income, and so forth" (Shneiderman 2000, 89).

Public Policy Statements on Gender Equity for the Information Infrastructure: Public policy statements with respect to gender equity for the information infrastructure have emphasized the need for women to become integrally involved as users, creators, and policymakers. For instance, the final *Beijing Declaration and Platform for Action* from the Fourth World Conference on Women (to be discussed in more detail in the next chapter) reiterated the need for women, especially in developing countries, to enhance their skills, knowledge, and access to information technology. [6]

In the United States, gender equity recommendations for the National Information Infrastructure (NII) have not been promoted. For instance, *A Nation of Opportunity*, the final report of the United States Advisory Council on the National Information Infrastructure, does not mention gender as a factor affecting access (U.S. Advisory Council on the National Information Infrastructure 1996).

Canada has been one of the few countries to consider the importance of gender equity to the information infrastructure in their public policy deliberations. The Information Highway Advisory Council's (IHAC) final report recognized that gender and social barriers need to be removed to ensure equitable and universal access to the information infrastructure: "Women's issues and concerns…must be addressed. Some of these, such as safety, privacy, and security, could be largely addressed by early implementation of related recommenda-tions…women have to be able to use the Information Highway and contribute to the content carried. The government can raise the awareness of content and hardware providers and can also implement public awareness campaigns tar-geted to women" (Information Highway Advisory Council 1997).

With the release of *Building the Information Society*, the federal government's response to the IHAC final report, the government reiterated their commitment to ensuring universal access to all, including an examination of gender as one factor affecting access. [7] However, as the public interest group the Coalition for Public Information (CPI) noted, IHAC's recommendations for gender equity were very weak, and did not address many of the gender issues recommenda-tions CPI detailed, including development of online gender issue information services and gender-sensitive educational software and training materials, and online harassment guidelines (Skrzeszewski and Cubberley, December 1995c). Although there have been considerable pressures for federal entities to apply gender-based analyses to issues surrounding universal access (Shade 1997b), to date no substantive recommendations or policies have emerged.

In Europe, the European Commission's ISPO (Information Society Project Office) has sponsored a project called OPUS — *Opening Up the Information Society to Women*. The main goals of the project were to develop a training manual and a CD-ROM on improving women's participation in the Information Society (www.iris-asbi.org/opus). In Britain, the Economic and Social Research Council (ESRC) has funded a seminar series, "Equal Opportunities Online," to explore the impact of gender relations upon the design and use of ICTs (www.regard.ac.uk)

LOOKING FARTHER INTO THE FUTURE

Proposed reforms to get women to use and participate on the Internet have included attention to both the technical and the social infrastructures. Forging beyond the technical requirements of getting online and coupling this with an exploration of the social factors that can affect access is of utmost impor-tance. Although many of these barriers to access are not just gender specific but encompass a diverse citizenry, they do shed light on the intricacies and nuances of factors such as network literacy, social facilitation, and governance.

So far, this chapter has considered the overall workplace dynamics, includ-ing women's relationship to paid work, telework, and the redomestication of work. Issues of substance and content have focused on the development of online gender information services, the definition of essential services, and the debates surrounding privacy, sexual harassment and pornography. Moreover the development of feminist, user-centered design was considered, taking into account the problem of essentializing (and often trivializing) design.

The issues surrounding universal access are complex and multifaceted, existing as they do in a rapidly changing and technologically converging and competitive environment. Given the swiftly changing nature of the Internet, policies and reforms need to be fluid to meet the demands of a diverse population. Funding and support mechanisms to allow access to basic network services for the citizenry are necessary to mitigate the distinctions between the information have's and have-nots, or what is now referred to as the digital divide.

Although reforms to make universal access a priority are important, it is also important to critically assess the role of the Internet in our everyday lives. In an era of deficit reduction, what should the priorities be for educational and social service delivery? Do we fund Internet access in every K–12 school while slashing teachers salaries, increasing class size, and halting physical reconstruction of school buildings (Shade 1999)? Do we spend millions creating community access points for the Internet while eviscerating publicly funded services such as healthcare/Medicare, welfare, and daycare? Do we streamline government service delivery by digitizing content while laying off public service workers under the guise of efficiency? Does our workforce need to be re-engineered by the creation of virtual corporations and workplaces?

As well as critically analyzing values implicit in our new digital economy and communities, we must also realize that the Internet can be a locus of activism, consciousness-raising, and enhancement. Examples of using computer networking as a form of resistance and defiance provide useful case histories, as documented in Chapter 2.

Another key area is the cognizance of the subtle shift in how women's traditional domestic roles are being redefined due to the ability of networked communications to blur the demarcation between the private and public spheres; the area of telework and the redomestication of work is only one such example.

It is important to differentiate between the placement of informational appliances in the home sponsored by a cornucopia of information providers with attendant visions of interactive home shopping on demand and the public interest which takes as its focus the citizen and everyday community needs. The business prospects and the managerial implications of the information superhighway, the highly touted delivery system of the near future that will bring entertainment and communication services to both the domestic and business sector, have been the focus, and public interest issues have been sideswiped in the rush to commercialize and privatize the information infrastructure within a competitive environment. Citizens have been reduced to mere passive consumers of the products that the telephone, cable, and television industries want to propagate.

There is a prevailing social discourse surrounding the Internet, as reflected in advertising, the media, and in some commercial applications, that situates networking technology within the domestic sphere. A cursory examination of current advertising in popular computing magazines reveals the same theme: white, nuclear families gathered around the new electronic highway hearth. Indeed, the discursive strategies used to debate the new interactive technologies are surprisingly similar to those used to discuss the introduction of television into the post-war economy and new suburban landscape, where television came to be seen as a window onto the world, and spectatorship became privatized and domesticated. It was also a time for the re-entrenchment of women into the domestic arena, the proliferation of the nuclear family sensibility amidst cold-war rhetoric, and the burgeoning spread of single-family homes in the new Levittowns (Spigel 1992).

Whether or not new competitive trials of interactive services will increase women's access to a variety of networked services is unknown at this time. In considering the incursion of various innovative networked technologies into the home, we must ask how will family structure and community life be shaped? In the design of these so-called smart homes, are women considered a relevant social group by the designers, architects, and technologists (Berg, 1994a & 1994b)? Will the coming digitization of the home, like its industrial predecessor that Cowan (1983) so vividly described, create yet again more work for mother? A recent advertisement shows a photograph of a beaming, white, middle-class women in her kitchen, perched over her NetVista X40i and her glass of orange juice, with the copy reading "My new NetVista X40i fits my life and my décor."

Trials of interactive technology have been hampered by technical and logistical difficulties, and predictions of market penetration are not as rosy as the telephone and cable firms envisioned, with the technology "unlikely to reach even one-quarter of U.S. households until well after the year 2000" (Cauley 1995, A1). What is disturbing is how, in the general current discourse of these user-centered trials, the users of the technologies are considered consumers of the services that will be delivered to their homes, rather than active and inquisitive citizens who might use the technologies for personal empowerment or edification.

Furthermore, even if the resources conducive to personal edification are available, it takes leisure time to be able to use them. This issue has a gender component. Economist Juliet Schor (1992) has documented an average increase in working hours (the equivalent of an extra month a year, or 163 hours) for most full-time employed people.

Women, because of their prescribed gender roles, generally have even less leisure time than men. This concomitant erosion of leisure is particularly acute for working mothers, with estimates of time spent on domestic responsibilities ranging from 25 to 45 hours per week. Reiterating Schor's observations, and commenting on figures culled from Statistics Canada, Gadd reports that "even when employed, women perform the lion's share of child care and household chores. Employed women spend about 2.5 hours a day cooking, cleaning and washing clothes, an hour a day more than employed men. Employed married mothers spend an average of 2.2 hours a day on child-care activities, double the time their spouses spend" (Gadd 1995, A5). We need to ask ourselves whether the Internet will lead to an expansion of everyday tasks, raise standards of work or efficiency, or increase or decrease our leisure time.

It could be that the Internet will become yet another example of how women's use and appropriation of a communications technology changes its original trajectory—in this case, the current commercial rhetoric which essentially espouses a one-way flow of information and pitches consumer marketplaces. As documented in Chapter 2, the unintended consequences of women's use of technology are vividly illustrated by the social history of the telephone: "...from the first decades of the twentieth century, women used the telephone, and used it often, to pursue what they, rather than men, wanted: conversation" (Fischer 1992, 233). Indeed, the aforementioned Pew Internet & American Life Project claims that women's use of the Internet "has reshaped America's social landscape because women have used email to enrich their important relationships and enlarge their networks" (2000). E-mail is akin to the use of the telephone for "kin-keeping," as documented in Moyal's study on women's use of the telephone.

CONCLUSION

Feminist perspectives on technology stress the social context of technology where the importance of the various and heterogeneous social factors in the shaping of technological design, change, and diffusion, and the interrelatedness of the work, lives, and status of the producers and consumers are explored. Judy Wajcman (1991), in demonstrating that political choices are integral in the implementation and design of technologies, urges an analysis at both the design level of specific technologies and at its location in the public and private spheres. How will the Internet affect women's status, and what is its relevance for women? More studies need to be conducted, and also examining issues of access in a variety of contexts—the domestic to the workplace—to issues of content creation, design, public policy and gender equity.

Ursula Franklin has written that women's greatest contribution to the current technological landscape lies in their potential to change the present structure by "understanding, critiquing, and changing the very parameters that have kept women away from technology" (Franklin 1990, 104). Focusing on the gendered implications of access not only allows us to widen the spectrum of social actors who participate and partake in the information infrastructure, but can also lead us to consider some of the unanticipated consequences of new technologies.

However, as has been mentioned before, essentialist analyses, which assert that there are fixed and unified opposed male and female natures, can often be too limiting. The International Development Research Centre (IDRC) Gender and Information Working Group recommends that policy recommendations should account for several factors, including accounting for the information needs of both men and women to "foster an understanding of the mutual benefits to be gained by society" (IDRC 1995, 279) and that approaches should be participatory, encouraging community-wide participation in the design and management of initiatives.

As our public policies for the Internet develop, it is imperative that a more inclusive decision-making process be created (Kramarae 1997). So far, policy-making in North America has engaged an elite of technologists, bureaucrats, industry, and academics. Public interest groups have had only minimal presence, and in most instances, their presence came about after vociferous protests. Therefore, the needs of a diverse citizenry, including women, seniors, racial and ethnic minorities, the disabled, and those marginally employed, need to be included in these important debates. The next chapter will look at the situation of women in developing countries, whose access issues vary considerably from women in the North.

NOTES

1   One of the ways to make the computer culture more inclusive is to recognize and accept the different learning and stylistic paths men and women take in their com-

puter programming and design—what Papert and Turkle dubbed "epistemological pluralism" (1990). The move towards object-oriented programming, and a change in computer interface design from highly abstract UNIX line commands to WYSIWYG (what you see is what you get) designs, featuring icons fashioned as familiar desktop objects such as trashcans and files (the Apple Macintosh version) to the Microsoft Windows design, is one way the computer industry has tried to reach out to a wider audience. The new trend towards social interfaces for PCs featuring real-life scenarios (living rooms, offices, town centers) is another way the computer industry has attempted to humanize the computer, particularly as the computer becomes more domesticated. See Sherry Turkle and Seymour Papert, Autumn 1990, "Epistemological pluralism: Styles and voices within the computer culture", *Signs* 16:128–157.

2.   Mid-1990s demographic surveys on the Internet revealed that women's participation on the Internet was not as high as that of their male counterparts, but that their use was slowly increasing. O'Reilly and Associates, in their 1995 survey *Defining the Internet Opportunity*, put the U.S. Internet male-to-female ratio at 67% male and 33% female. See O'Reilly and Associates. October 1, 1995, URL: www.ora.com/research/users/charts/net-gender.html. GVU Center's 6th WWW User Survey in 1996 indicated that the gender ratio in North American was 31.4% female and 68.6% male, and that European users were also still predominantly male (80.2%). URL: www.gvu.gatech.edu/gvu/user_surveys/survey-10-1996/. A 1995 telephone survey by J. Katz and P. Spading revealed that longtime Internet users were more likely to be male (66%); new Internet users are slightly more likely to be male (55%), but that the gender gap was decreasing, as their 1996 survey revealed that female users were 46% of Internet users. See "Motives, Hurdles and Dropouts: Who Is on and off the Internet, and Why." *Communications of the ACM* 40 (April 1997): 97–102.

3.   A concise description of the Communications Decency Act can be found on pp. 136–8 in *Virtual States: The Internet and the Boundaries of the Nation-State* by Jerry Everard (New York: Routledge, 2000).

4.   June 1995 personal e-mail with Pat Nelson.

5.   Feminist-standpoint epistemologies, as advocated by Sandra Harding and Hilary Rose, for instance, claim that women are privileged epistemologically in that their historically underrepresented position in society produces more accurate and better accounts of the real world—an integration of "hand, brain, and heart," as Rose puts it. Harding questions whether there can be a single feminist standpoint when there is a multiplicity of women encompassing a variety of races, classes, and ethnicities, and she would have us embrace a "successor science." In response to this proposal, Haraway suggests that the politics of the partial perspective, a notion of objectivity that "privileges contestation, deconstruction, passionate construction, webbed connections, and hope for transformation of systems of knowledge and ways of seeing" (Haraway 1991, 191–2), would be an apt maneuver. See Sandra Harding, *The Science*

Question in Feminism (Ithaca, NY: Cornell University Press, 1986); Donna Haraway, pp. 183–201 in "Situated Knowledges: The Science Question in Feminism and the Privilege of Partial Perspective" in *Simians, Cyborgs, and Women: The Reinvention of Nature* (London: Routledge, 1991); Hilary Rose, "Hand, Brain, and Heart: A Feminist Epistemology for the Natural Sciences." *Signs* 9 (Fall 1983): 73–90.

6. Beijing Declaration and Platform for Action, 1995. URL: women.usia.gov/usia/beijpg.htm.

7. Information Highway Advisory Council, *Connection, Community, Content: The Challenge of the Information Highway*, Ottawa: Minister of Supply and Services Canada, September 1995; and *Building the Information Society: Moving Canada into the 21st Century*, May 1996, URL: strategis.ic.gc.ca/ihac

6 | BEYOND BEIJING
Strategies for the Next Wave

> ICTs (information and communication technologies) will be one of—if not the—major development issues of the coming decades. If women are not actively present at all levels, we will see new forms of marginalization that could undermine other advances made by women in the 20th century. This implies a crucial challenge to women to take on these issues (*Information and Communication Technologies: A Women's Agenda*, APC Women's Networking Support Programme, 1999. URL: www.gn.apc.org/apcwomen/resources/policy/women-rights.html).

As the last chapter showed, the issue of access to the Internet is a vital concern, encompassing both the social and the technical infrastructure. The risk is that, even as more people get online, the digital divide will increase so that more women, particularly those in the lower socioeconomic bracket, will be left behind. This chapter looks at the issue of access to the Internet for women in developing countries, where access barriers are more challenging and present different ethical conundrums.

When looking at the developing world, it is necessary to consider information and communication technologies (ICTs) as enabling technologies that can support many social, cultural, political, and economic activities within communities (Madon 2000). Potential benefits of ICTs for development include strengthening local capacities; expanding the knowledge base; creating new opportunities for employment and entrepreneurship; enhancing service delivery for citizens and communities; and improving the cost-effectiveness of health care delivery and service (Crede and Mansell 1998).

Despite the significant activities surrounding the Global Information Infrastructure (GII) [1], issues surrounding gender have not been directly addressed. Now we find GII arguments being subsumed by the rhetoric surrounding the creation of a knowledge-based economy/society, which emphasizes the interrelationship between ICTs (information and communication technologies) in knowledge and economic development. Tellingly, the OECD countries are just concerned with the KBE whereby knowledge is the "driver of productivity and economic growth" (OECD 1997). Knowledge becomes an entity to be commercialized and exploited. For instance, Industry Canada proclaims that the prime characteristic of the KBE is its "ability to generate and use knowledge—to innovate" which is "not only a determinant of wealth...but also the basis of comparative advantage. Knowledge is the fundamental means to improving the efficiency of production and distribution processes, improving the quality and quantity of products, and increasing the choice of products and services for consumers and producers" (1997).

The KBE brings with it a whole new vocabulary replete with new metaphors and expressions. We're in an "attention economy" where consumers can quickly "churn" on you even though we're in a "disintermediated" environment of "friction free" capitalism where intellectual or "human capital" reigns supreme. Where is your "mindshare" going tomorrow? (See Hotwired's *Encyclopedia of the New Economy.*)

For Canada and other OECD countries, neoliberal policies (globalization, telecommunications deregulation, privatization, liberalization, and competition) are enabling factors in the transition to a knowledge-based economy. These are the same strategies that are allowing the consolidation of advertisment-based commercial media (Eisenstein 1998; Schiller 1999).

The literature on women and development is vast (Ward and Pyle 2000) and resources on women, development, and technology are also considerable (Everts 1998). A burgeoning literature exists on women, development, and information and communication technologies (ICTs) (Sweetman 1999). Global strategies that encourage gender equity to the Internet have been conceptualized and in some cases initiated, due to the initial and follow-up efforts of the Fourth World Conference on Women held in Beijing in 1995. But, as the case study of integrating gender concerns into the first Global Knowledge conference will show, gender concerns are shunted to the side. This chapter examines issue of access to the Internet for women in developing countries, describes Section J in the *Beijing Platform for Action*, and looks at various Beijing +5 activities, whose goals have been to assess the efficacy of Section J.

WHOSE GLOBAL KNOWLEDGE?: GETTING WOMEN INTO GK97

In June 1997 the World Bank, with the participation of the Government of Canada through the Canadian International Development Agency (CIDA), organized one of the first conferences to focus on the use of information and communication technologies for developing countries: *Global Knowledge 97: Knowledge for Development in the Information Age*. When word of the conference and invited keynote speakers and participants was disseminated on the Internet in March, many were outraged at the dearth of women. A group was quickly organized in Toronto to lobby the conference organizers to get more women involved in the conference as speakers and participants. The Ad Hoc Committee for Women at GK97 consisted of ABANTU for Development, Black and White Communications, the APC Women's Network and Web Networks (an affiliate of the Association for Progressive Communications), the Canadian Committee of UNIFEM (United Nations Development Fund for Women), the Metro Action Committee for Women (METRAC) and the Women in Global Science and Technology Project (WIGSAT), as well as several independent Internet designers (Post Industrial Design, coolwomen.com).

Three major initiatives were developed by the Ad Hoc Committee: (1) a global e-mail campaign to nominate women participants; (2) the establishment of the Ad Hoc Committee as an Associate Sponsor of the conference; and (3) a presentation on the gendered implications of ICT policy, design, and implementation at a women's breakfast attended by the Presidents of CIDA and the World Bank. A GK97 Virtual Conferencing Project was also initiated, and one of the online conferences was devoted to discussions of gender and ICT policy, with the primary goal to develop a gender and IT platform to present at the conference (the "Canon"). [2]

Sophia Huyer, in assessing the efficacy of the e-mail campaign, acknowledges that the task of genderizing GK97 was successful: "many more women attended as presenters, speakers and participants than would have if there had not been a campaign...many more women from different regions attended...the impact and range of the campaign showed the organizing institutions that there actually exists a pool of women who are experts in this area." However, she also acknowledges that "women were still missing from some of the higher-powered discussions…and from the deal making that went on informally throughout the conference" (1999, 125).

The second Global Knowledge Conference was held in Kuala Lumpur in March 2000, hosted by the Malaysian government and the Global Knowledge Partnership, an informal coalition of intergovernmental and development organizations, businesses, and NGOs. Gender issues received more prominent coverage than at the first conference. The Global Knowledge Women's Forum was organized by the National Council of Women's Organizations of Malaysia (NCWOM) and the United Nations Development Programme (UNDP) in order to ensure that a gender perspective was integrated into the conference and the Action Plan. Two initiatives were included in the Action Plan. The first, the Gender and ICT Replication and Learning Fund, aims to promote an exchange of initiatives on gender equity and women's empowerment using ICTs. The second, the Support to Women Entrepreneurs, provides incentives either through ICT businesses or through online mentoring and financing (see *Transcending the Gender Information Divide* 2000).

ACCESS ISSUES FOR WOMEN IN DEVELOPING COUNTRIES

> As it is at present, the Internet reflects the perceptions of Northern society that Southern women are brown, backward, and ignorant. An alternative, kinder depiction of them, which is also widespread, is that they are victims of their cultural heritage. Is being exposed to such images of themselves going to help Southern women by encouraging them to fight in dignity and self-respect, or will it further erode their confidence in their fast-changing environment? (Annapurna Mamidipudi, in Gajjala and Mamidipudi 1999, 14).

Access to the Internet encompasses not just physical and technical mechanisms, but a myriad of economic and social factors. Public policy discussions on access, including those surrounding the GII, mainly discuss the technical barriers towards access, for instance, the hardware and software to support communication, resource discovery tools, and issues surrounding standards, interoperability, privacy, security, and intellectual property protection.

It is assumed, and advocated, that private industry will take the lead in developing and building the diverse technical components that go into making information highways. Furthermore, advances in developing and deploying the GII/KBE will depend on deregulation and the fostering of a competitive marketplace. Public policies, both national and international, have been therefore promoting the removal of regulatory, trade and policy barriers that could possibly impede competition and limit widespread deployment of the GII/KBE. In this current fiscal and policy climate, the marketplace is presumed to drive both availability and affordability (and hence, meet universal access goals). [3]

Access to the Hardware and Software to Support Communications. In developing countries, the use of networking technologies as a tool for international dialogue and information exchange is increasing, but their development should allow for equal participation by women and men of all incomes, languages, and locations.

The Panos Institute examined Internet access in developing countries in the South, and concluded that the gap between developing and developed countries is severe. Internet technology is both scarcer and more expensive in the South. An information gap already exists, in that at least 80% of the world's population still lack the most basic telecommunications: "The information revolution has only reached a few universities, companies, journalists, researchers and governments in developing countries. There is a danger of a new information elitism which excludes the majority of the world's population" (1996). For example, Africa has 12% of the world's population, but a mere 2% of its telephone lines, with most of the phone lines in urban areas (Marcelle 1997, 2). [4]

Obstacles to Internet access include physical factors such as inadequate telephone and electricity supply; the lack of computer and other technical equipment; outdated or defunct software packages; and the high cost of general telecommunications infrastructure, such as telephone connections and equipment. The United States, the European Union, and Japan dominate global telecommunications, sharing 74% of world revenues from telecommunications (Crede and Mansell 1998).

Social factors include the lack of adequate training and the staff necessary to support such training; and the absence of information and resources in local languages. For instance, Africa has several thousand languages and dialects, but very few technological products (Marcelle 2000, 58).

Whose Knowledge? Vandana Shiva (1997) warns of an encroaching monoculture, where Western knowledge and ideals are foisted on countries in the South. She worries, appropriately enough, that the intellectual commons will be driven by market forces and global trade institutions such as the International Monetary Fund, the World Bank, and the World Trade Organization. This is especially at issue in developing countries, where women are often the depositories of local knowledge, particularly in the areas of agriculture and indigenous medicine. Although such traditional knowledge cannot be patented, it can be transferred to developed countries where market-driven patent systems are the norm.

POST-BEIJING ACTIONS

The final *Beijing Declaration and Platform for Action* from the Fourth World Conference on Women reiterated the need for women, especially in developing countries, to enhance their skills, knowledge and access to information technology. Its goal was the acceleration of strategies aimed at "removing all the obstacles to women's active participation in all spheres of public and private life through a full and equal share in economic, social, cultural and political decision-making" (*Beijing Declaration and Platform for Action* Item 237, 1995).

Although the right to communicate was an underlying premise of many of the objectives, specific objectives focused on women and the media. Strategic objective J.1 of the *Beijing Declaration* identified the need to "increase the participation and access of women to expression and decision-making in and through the media and new technologies of communication." A series of actions called on government, NGOs, the media, and private industry to encourage and recognize women's electronic networks; to promote and develop educational and training programs for women in new communication technologies; to encourage the use of computer networking as a means towards strengthening women's participation in democratic processes; and as a means of encouraging alternative media that promotes women's voices. (See URL: www.un.org/womenwatch/daw/beijing/platform/)

On 26–28 June 1996, an Expert Workshop on Global Information through Computer Networking Technology in the Follow-up to the Fourth World Conference on Women (FWCW) was held at the United Nations headquarters in New York. The Workshop, sponsored by the UN Division for the Advancement of Women (DAW), the UN Development Fund for Women (UNIFEM) and the International Research and Training Institute for the Advancement of Women (INSTRAW), looked at how to facilitate global information exchange for monitoring the implementation of the *Beijing Platform for Action* through the use of computer networking technology.

The Expert Workshop devised a series of recommendations towards implementing a World Wide Web site, WomenWatch. The workshop also looked
at issues surrounding access and training issues; best practices with global
networking for women; and principles for cooperation between NGOs
and the United Nations in the design and implementation of the WomenWatch
project.

In conceptualizing WomenWatch, several salient questions arose:
(1) How can an Internet site mobilize, build coalitions, and offer a space
to share experiences and lessons learned?; (2) How can women in the South
play an active role as both producers and creators of content on the Internet?;
(3) What partnerships can be developed to increase women's presence on
the Internet, particularly in the South?; and (4) How can a Web site "ensure
that the commitments made in Beijing become reality for women every-
where?" (WomenWatch, 1996).

Advocacy and mobilization through electronic communications via
three related activities was considered: (1) providing vital information
resources, (2) serving as an organizing tool and (3) facilitating outreach
activities. Partnering was another important activity identified. This would
include advocacy and mobilization with: diverse media—mainstream and
alternative; the private sector; nongovernmental organizations (NGOs);
diverse women's organizations; the United Nations; various governments;
the academic community; specialized libraries; and new users.

Another important area of action which WomenWatch identified
was to influence telecommunications policy on issues surrounding access,
security, privacy and intellectual property issues:

> It was recommended that the WomenWatch initiative play a proactive role
> in the telecommunications and information policy process within the United
> Nations system and vis a vis other multilateral and regional organizations,
> to ensure that gender considerations become and remain an integral part of
> those discussions and decisions. This would include a voice with groups like
> the World Bank, the International Telecommunications Union, European
> Union and the United Nations Commission for Science and Technology
> (ibid.).

Other areas identified by WomenWatch included the development
of a hospitable online environment for women, the development of software,
and the conceptualization of a broad World Wide Web strategy regarding
interoperability, interactivity, navigation tools, and language.

In 1999–2000, various activities centred on assessing the implementation of the *Beijing Platform for Action* through the Beijing+5 Review. Various regional NGO reports on women and media from Asia, Latin America, and the European Union were written. In addition, a Women and Media six-week online discussion was conducted, a joint effort of WomenWatch and WomenAction 2000 (a network of national, regional, and international organizations with a focus on Section J of the *Beijing Platform for Action*). The forum assessed whether the Section J objectives had been met. Weekly themes looked at women and the information society, the use of new communication technologies, media portrayal of women, women's networks and media, freedom of expression, and the social responsibility of the media. European, Canadian and U.S. participation was the highest; but women and women's groups from Asia and the Pacific, Africa and Latin America participated as well. The main concerns of the regional reports and the online forum included the following:

- Many of the concerns raised in Section J still remain, while new ones have emerged.
- Some progress has been made in implementing portions of Section J, due to the sustained monitoring, networking, and lobbying activities of women's organizations and media watch groups.
- The increasing concentration of media industries and the monopolization of culture and information industries are huge issues. Section J did not cover "the economic and political realities within which transnational media corporations perpetuate inequalities and inequities...and women's vulnerabilities, as traditional keepers of indigenous knowledge within this environment are not acknowledged. Women are concerned with the absence of analysis of the globalisation of media, particularly mergers of transnational media corporations and changes in media ownership at national levels that have a bearing on media content and intent" (WomenAction 2000, *Alternative Assessment*).
- A special concern is the decline of public broadcasting media.
- The increasing deregulation of communication industries (e.g., the 1996 Telecommunications Act in the U.S.), and OECD influences on the MAI, WTO accords, etc., need to be looked at critically.
- Women still lack access to basic infrastructure such as electricity and phone lines.

- There has been little commitment from public or private institutions to address women's situations.
- The "targeting of women and girls through pornography, sex tourism, and the sex trade has been exacerbated by electronic communications. The sex trade can now use websites for their portrayal of women, with the capacity to reach many more people. In the U.N. European region, the result is an increasing number of women-hating websites; this, and the use of flaming and stalking, tend to dissuade women in these parts of the globe from using the Net" (WomenAction 2000, *Alternative Assessment*).

TOWARDS GENDER EQUITY

ICTs offer new forms of communication that may enable women to break through their often isolated social situation. They also create new opportunities of employment for women in jobs that require new skills. However, the technologies themselves will not achieve this. The full use of opportunities that are in principle created by the deployment of ICTs will depend upon social variables such as cultural factors, class and age. Robust policies are needed for the ICTs to have a beneficial impact on women's lives" (Cees J. Hamelink 1999).

As was mentioned in the last chapter, policy recommendations towards achieving gender equity should account for several factors, including the information needs of both men and women, participatory approaches, and community-wide consultation in the design and management of initiatives. To ensure gender equity to the Internet for those in developing countries, public policy formulations on universal access should consider the following issues: access to user-centered design, training and education, and employment and workplace issues.

Access to User-Centered Design. Within the developing world, the promotion of women's indigenous knowledge and skills is vital. This includes allowing women to identify the information that they want and need, and assisting them in determining the appropriate delivery mechanism (e.g., a combination of online and offline modes). Involving women in the design is crucial. Moreover, communication systems must be designed to include locally appropriate tools, including radio, solar energy, video, and tape recorders (APC WNSP 1999). Repackaging information (e.g., from hard copy to electronic, and vice versa) is a way of making sure that it can be disseminated in as many forms as possible. Technical means to guarantee the privacy and integrity of information must be encouraged.

The role of telecenters is of particular importance in developing countries. Telecenters are one or more telephone lines in a single location, with workers present to provide basic assistance for placing calls. Some telecenters have expanded to include access to fax machines, photocopiers, e-mail, and the Web. Schreiner's (1999) study of the impact on women of a telecenter in Bamshela, South Africa, concluded that the women used the telecenter to telephone friends and relatives. The telecenter was particularly useful for women maintaining ties with family members working as migrant workers, but the cost was prohibitive for many.

Training and Education. Funding mechanisms for the creation of content by women and women's groups, and the provision of training programs in basic and advanced networking skills should be encouraged. To expand upon the recommendation of the *Beijing Platform for Action*, training and assistance in direct use of the Internet in developing countries needs to be broadened. Training should be culture and gender sensitive.

Likewise, the coordination of national and international efforts, particularly with respect to initiatives that provide training, access to hardware and software, and mentoring opportunities, within the context of the GII/KBE should be expedited.

Technology needs to be created and used in an enabling environment. Martha Nussbaum argues that international political and economic thought must be sensitive to gender justice, and focus on human capabilities, "that is, what people are actually able to do and to be—in a way informed by an intuitive idea of a life that is worthy of the dignity of the human being" (2000, 5).

Employment and Workplace Issues. Networked technology is dramatically changing work and the high-tech global economy. At the same time that it is creating innovative opportunities for cost-efficient collaborative endeavors, it is also leading to both de-skilling and job losses in many industries and countries (Klein 2000).

The effect of a knowledge-based economy on developing countries needs to be addressed and assessed. How can women's indigenous knowledge be transferred to this milieu? Women from developing countries are underrepresented in the use of new information and communication technologies, but ironically they tend to comprise the workforce that produces computer components, often in unhealthy working conditions (Ong, 1987). Is such work always exploitive? Swasti Mitter and Sheila Rowbotham comment that, in some cases, "economic power, autonomy and the chance to escape the tyrannies of traditional societies" (1997, 17) are achieved. Mitter (2000) also comments that in India, the software industry is women-friendly, compared with other types of engineering, with women comprising nearly 20% of programmers. However, women are concentrated in the low-value end of the software industry, rather than in more challenging technical and managerial positions. This is because "the child-bearing role of women often makes it necessary for them to settle for occupations and tasks that are less challenging. This is particularly so as working hours are generally geared to male needs" (ibid.) [5]

Using the Internet to market indigenous crafts has been one way that women in developing countries have achieved a level of economic self-sufficiency. However, challenging the status quo can often lead to tensions. Consider the Rupunni Weaver's Society. This is a women's weavers organization, consisting of 300 women from the Wapishana and Macushi tribes in Lethem, Guyana, who created a website to market their handmade hammocks. This e-commerce venture created tensions with the male regional leadership, which took over the weavers organization. Frustrated, the woman creator of the website quit (Romero 2000 and www.gol.net.gy/rweavers).

Policy Formulation. Women must be integrally involved in policy formulation on Internet projects that affect their local communities. Gender-based assessments on the impact of the Internet on women and their communities are crucial.

CAUTIONS ARE IN ORDER

We need to look critically at arguments which posit that if women in developing countries do not leapfrog over technological developments that they have missed, they will then find themselves further marginalized. The United Nations Commission on Science and Technology for Development cautioned that ICTs "do not offer a panacea for social and economic development. There are risks of unemployment and social and economic dislocation...however, on the basis of the evidence, it is apparent that the risks of failing to participate in the ICT revolution are enormous. Failure to give priority to ICT strategies that enable developing countries and countries in transition both to develop their national infrastructures and to join the GII will exacerbate the gap between rich and poor" (UNCSTD, 1997, 17).

The creeping corporatization of Internet development in developing countries also needs to be looked at critically. For instance, The Bill and Melinda Gates Foundation (along with the University of Southern California's Annenberg School of Communication) has lent support to Women Connect!, a million-dollar project whose aim is to assist NGOs in providing connectivity to rural women working in issues in health and environmental protection, particularly in Africa (www.women-connect.org). Although this is a laudable goal, one must ask when, and how, social accountability and sustainability should be highlighted.

It is important to look at what works. There is an extensive literature on the impact of ICTs on women in Africa (Rathgeber and Adera 2000). One tangible example is Women'sNet (www.womensnet.org.za), an initiative of SANGONet, an ISP (Internet Service Provider) in Johannesburg, South Africa. Women'sNet was launched in 1998 in collaboration with the Commission on Gender Equality, with the aim of increasing South African women's capacity to use the Internet by the development of relevant content. The first topics presented were the prevention of violence against women in South Africa and the situation of women in small, medium-sized, and microenterprises. Women'sNet also tries to enhance the ability of lawmakers, policymakers, and civil society to redress the unequal status of women in South African society. For instance, information on government announcements and how to submit parliamentary committee motions has been disseminated (Opoku-Mensah 2000). Other information sites include the Women and Human Rights site (womensnet.org.za/human-rights), which provides human rights information, the Women and Enterprise site, and an information section on health and HIV/AIDS information.

One reason Women'sNet is successful is that it considers local needs and concerns. As Gillian Marcelle points out, in African society, a key element of female poverty is time poverty: "women work longer hours than men…and their workload, derived from simultaneously carrying out multiple roles, imposes severe time burdens and harsh trade-offs, with important economic and welfare costs. Balancing competing time uses in a framework of almost total inelasticity of time allocation presents a particular challenge to reducing poverty in Africa" (2000, 73).

So far, women in developing countries have not been the targets of consumer-oriented information on the Internet. Indeed, e-commerce initiatives where women market their products on the Internet appear to appeal to those in the North. Fair trade organizations have been active in marketing various artisanal crafts while handling secure payments and other logistical concerns. In India, the Self-Employed Women's Association (SEWA) artisan support program works with numerous women artisans from the Banaskantha district, north of Gujarat (UNIFEM 2000, and www.banascraft.org/banascraft/sewa.htm). Here, women are able to use the Internet to exercise a degree of economic self-sufficiency, playing the role of producer, rather than consumer.

The ongoing tension between producers and consumers of the Internet is illustrated in two competing moral visions of the Internet—that of e-commerce versus e-commons, which will be the topic of the concluding chapter.

NOTES

1.   U.S. Vice President Al Gore has been instrumental in promoting the GII. He exults that the GII "will forever change the way citizens around the world live, learn, work and communicate...The GII is a historic undertaking. It is strengthened by participation, bolstered by openness, and fortified by strong nations and talented people pursuing dreams of a better tomorrow." (See *Global Issues, Electronic Journals of the U.S. Information Agency*, Volume 1 (September 1996), URL: www.usia.gov/journals/itgic/0996/ijge/ijge0996.htm). To promote the U.S. vision of the GII, the telecommunications sector has been liberalized and there is a movement to work with international governments to eliminate obstacles that stand in the way of communication and information industries competing in foreign markets. Critics (non-U.S. citizens) have remarked that GII stands for "Global Information Invasion," see *Borders in Cyberspace: Information Policy and the Global Information Infrastructure*, edited Brian Kahin and Charles Nesson, Cambridge, MA: The MIT Press, 1997.

2.   *The Canon on Gender, Partnerships, and ICT Development* was presented at the Global Knowledge '97 Conference, June 22–25, Toronto, Canada. Its three basic aims were: (1) ensuring that both men and women have input into the design of ICT systems; (2) assessing the differential impacts on men and women of the effects of ICT development programs; (3) evaluating ICT development to take into account the distinct situations and resources of men and women. Three critical priority actions identified were: (1) incorporating gender analysis into all technological and scientific policy research; (2) developing and financing follow-up and evaluation mechanisms to identify and measure the impact of ICT initiatives on women's communication needs; and (3) developing partnerships with women and their organizations to develop information systems specific to the needs of women.

3.   For a look at international public policy pronouncements on universal access, see "Universal Access: The Next Killer App," by L.R. Shade, a paper produced for *Defining and Maintaining Universal Access to Basic Network Services: Canadian Directions in an International Context*. Invitational Workshop sponsored by Industry Canada and the Faculty of Information Studies, Toronto, 14–16 March 1996. URL: www.fis.utoronto.ca/research/iprp/ua.

4.   The International Telecommunications Union (ITU 1998, 1999) has detailed the disparity in telecommunications access in developing countries. See www.itu.org

5.   This concern is also relevant to developed countries. Paulina Borsook, commenting on the situation in Silicon Valley, remarks that the overall tendency in high-tech firms is subtly antiwoman, which is why women tend to leave the field at twice the rate of men. "Whether it's nature or nurture, women tend to get stuck with more of the domestic management of a household. Caring for and about children runs right into the demands of a 70- to 80-hour workweek and tends to feed the suspicion that such weeks are not necessarily worth having. Yet on paper, in theory, it would appear that in the telecommuting, free-agenting roaring digital '90s, women could work well at home, programming at odd hours between tending to toddler upchuck and carpooling daughters to aikido practice" (Borsook 2000, 160).

7 | CONCLUSION

Citizens or Consumers?

These cyberdialogues exist between the cracks of mainstream news reporting and people's everyday lives. Such communication allows for diasporic publics to connect with one another and initiates new alliances with people "outside" one's immediate geographical region. These disparate and dispersed communications can be used for change and mobilize struggles for peace, equality, and a healthy environment" (Zillah Eisenstein 1998, 168).

Gender and Community in the Social Construction of the Internet has provided a critical feminist and political–economic perspective on the current trajectory of the Internet and the salient qualities of digital capitalism, particularly as women are targeted by corporations as a viable commercial market. It has highlighted the tensions between these corporate strategies (the *feminization* of the Internet) and the use of the Internet by women for activism (*feminist uses* of the Internet).

This dichotomy is not new. Through looking at the work of a variety of feminist scholars who have examined the gendering of other communication technologies (the telephone, the radio, and television), we have been able to see how women have been gendered through the social practices that have been promoted by industry, as well as through the uses adopted by women and their communities.

Women as active agents in the construction of the Internet is a theme woven throughout this book, which has provided an overview—albeit cursory—of some of the diverse women's communities using the Internet for feminism, activism, and democracy. It is important to remember that the Internet is one tool among many to link the international women's movement. There are still barriers to access for many, based on both technical and social infrastructures. Kramer and Kramarae (2000) remind us that traditional forms of networking are still vital. They cite the example of press networks such as Fempress in Latin America; radio, as exemplified by the Feminist International Radio Endeavor (FIRE) and the Women's International News Gathering Service (WINGS); and ISIS International, which utilizes computers, the postal system, the telephone, and fax.

Innovative uses of the Internet have been designed by Les Penelopes in France and Amazon City Radio. Les Penelopes has a weekly interactive television program called "Cyberfemmes," with online discussions and background texts to elaborate on the content of their programs. Its primary goal is to relay the reflections, actions, and struggles of women around the world. Recent program topics have included the status of women in media, the history of transsexuality, global prostitution, globalization, educator Maria Montessori, and women in sports (www.penelopes.org). Amazon City Radio, "the voice of women on the Internet," is a web-based radio station for women, using the RealPlayer. It archives the reports of WINGS (the Women's International News Gathering Service), which provides news about women around the world (radio.amazoncity.com).

But however much we applaud feminist activity on the Internet, we must remember that the feminization of the Internet is continuing unabated. I am referring to the creation of consumer-oriented content aimed at women by media behemoths and a new breed of competitive entrepreneurs. Such activity shouldn't be terribly surprising; we have seen how, with the emergence of every new communications technology, women have been targeted as a specific audience demographic by industry, advertisers, and the media themselves. It is important, then, to consider the wider issues of political economy if the existing patterns of ownership, control, representation and creation of women's content on the Internet is to be understood and challenged. As WomenAction's *Alternative Assessment* reported, "ICTs are double-edged swords: [they are] so often owned by multinational corporations and/or outside of our own control, [that] we cannot adequately know what is going on at all times. There continues to be a deeper exclusion for those who do not have access, and the digital divide has gone from being a theoretical talking point to a reality since Beijing, while we have seen few programmes aimed at reducing the gap for women" (WomenAction 2000, *Alternative Assessment*).

What are the concerns about privacy and women's freedom of expression and right to pluralistic information sources in our hypercompetitive media environment? Margaret Gallagher rightly comments that a view of competition is radical when its starting point is human and cultural diversity rather than financial markets (Gallagher 1996).

We must constantly be aware of how the idea of the technological fix as panacea pervades not just the discourse, but the actions surrounding technology in society. For instance, Canadian Finance Minister Paul Martin's 1999 budget called for $1.8 billion over four years to promote the "creation, dissemination, and commercialization of knowledge." The spending would ensure that all Canadians have a chance "to learn and profit from the Internet." This includes money earmarked for smart community initiatives. As quoted in an online Reuters news article: "…one government official said that the experimental projects would allow police departments to electronically notify parents of a missing child…or enable social workers to use digital voicemail to stay in touch with homeless clients" (see "The Most Connected Nation?"). The absurdity of these purported applications defies explanation. Something is truly out of sync in our values when we consider digital devices more important than shelter.

It is necessary to adapt a critical stance when looking at the role of the Internet in our daily lives. How, in information society discourses, are we talked about? Are people addressed as citizens or consumers? Vincent Mosco contends that we need to critique debates about citizenship as they are inserted into the discourses of the information society and knowledge-based economy. Which citizenship are we talking about? There is citizenship "as a bundle of legal rights derived from the sovereign state and citizenship as democratic participation in a community" (Mosco 2000, 44).

Although it is important to present positive and glowing descriptions of the possibilities and outcomes of the Internet in order to stimulate and promote the activities of women, it is also important to keep in mind that constructive and socially redeeming policies need to be put into effect. This book has showed how such policymaking should work towards ameliorating the current gender disparity in access (technological and social) to the Internet, and towards providing practical solutions regarding issues surrounding work and employment, and privacy and security. Given the fast pace of the implementation of a global information infrastructure, it is urgent that women in developing countries have a voice in the development, dissemination, and deployment of the technology, and that women in developed countries assist them in these endeavors. Policies must promote women's participation in political debate and government decision-making.

THE RIGHT TO COMMUNICATE

One of the ways to institute policy is through various Right to Communicate measures. The Right to Communicate has received increasing focus with the 50th anniversary of the Universal Declaration of Human Rights. Many of these principles first received attention in the 1970s with the New World Communication and Information Order (NWCIO). With the development of a global information infrastructure, the Right to Communicate in a "Network Society" (Castells 1997) has become even more challenging in light of swiftly developing technologies, a deregulated telecommunications environment, neoliberal economic and social policies, and a diverse civil society. Moreover, the increase in non-governmental organizations and their use of information and communication technologies has increased the urgency to establish global recognition of the Right to Communicate. Although a universal Right to Communicate has yet to become a part of accepted international law, there are several initiatives promoting this discussion at a global level.

The Right to Communicate is recognized by many governments as a fundamental human right and includes notions of freedom of opinion, freedom of expression, freedom of information, and freedom of the press. The Right to Communicate has also been expanded to include the notion of universal access to information and communication technologies, with access seen as both a technical and a social infrastructure (Clement and Shade 2000); the right to public access; and public participation in both the means of communication and towards policymaking. Linguistic rights are also a feature of the Right to Communicate, as well as the sustenance of indigenous languages and culture.

Media concentration and globalization are some of the inhibiting factors to the Right to Communicate. Of particular concern are issues of the global commercial media market, and corporate concentration of media industries, which are primarily American in culture and ownership; and the dominance of global capitalism, exemplified by institutions such as the World Trade Organisation (WTO), the World Bank, and the International Monetary Fund. It is important to examine the commercialization of information, particularly tax-supported government information, and intellectual property balances between creators, publishers, and the public interest. Corporate accountability and public participation in consultations surrounding policy formation at local, national, and international levels are also necessary prerequisites to ensure that all the diverse parts of civil society can fully exercise the Right to Communicate.

FUTURE RESEARCH

As this book has chronicled, despite the frenzied moves to attract and keep women on the Internet in commercial-driven branded environments, many women, women's groups, and young girls have been creating energetic and vibrant web content. Net-based communication has facilitated the design, development, and diffusion of feminist content—whether it is political, informative, creative, or humorous. In fact, many of the strategies and styles originally developed by early net feminists have been appropriated by commercial entities, whether it be a funky graphics style, a feisty attitude of can-do, or linking content with chatrooms and conferencing.

But, despite Steve's, and Bill's, and Oprah's incursions onto the World Wide Web, despite synergistic couplings of television broadcasters and telecommunication giants, despite the rapid development of broadband applications (and the increasing domestic placement of high-speed modems), I believe that feminist communication, and the use of the Internet by a diverse range of civil society, will be sustained. This doesn't mean that we should sit by complacently as AOL creates yet another cyber-mall; or as public policy continues to emulate the libertarian ethos of so much of Internet culture. As digital capitalism creeps its way into our daily lives, we need to be more conscious than ever of preserving the public interest.

The public interest spirit that has guided the communications sphere has historically included the mandate towards universal service to telephone services and the creation of non-commercial public-affairs, arts and cultural programming services in radio and television broadcasting. With the advent and proliferation of new communications technologies, public interest media attributes have become more elusive and complex. The era of telecommunications deregulation has brought about an ambiguous relationship between public and private interests: privatized and commercial interests versus governmental regulation and control. Moreover, networked technologies are constantly evolving and being deployed in new social arenas which transcend constrained geographic boundaries.

As an area of academic study, the Internet is in its nascent phases. Terrific research has been conducted on the interstices of gender and the Internet. More needs to be done. Here are some suggestions, ideally to be theorized from a critical feminist perspective:

Governance: Issues of governance of the Internet need to be analyzed and studied. How will the Internet be run? Currently the ICANN (Internet Corporation for Assigned Names and Numbers), a nonprofit corporation, is acting as the Internet's central coordinating body. The 19-member Board of Directors provides decisions regarding domain name registries and disputes. Controversies have arisen over the composition of ICANN, mostly in terms of ensuring both global and nonprofit representation. Not as well discussed have been issues of gender representation. I suspect that this will become a pressing issue in the next few years.

Labor: How are women participating in Internet industries, whether in the more mainstream industry organizations, via new start-ups, or through entrepreneurship ventures? How are women participating in the lower echelons of the tech force?

Women in technopoles: New digital broadband districts, often referred to as "smart communities," are located in major urban centers. How are women participating and engaging in technopoles? Who is excluded from these high-tech communities? How is citizenship conceptualizied in technopoles?

Qualitative studies: Studies of the use and impact of the Internet in the domestic context, and its negotiation within diverse families would be fascinating (Lohan 2000). What is the social shaping of Internet technologies in households? How are consumption and use influenced by industry stakeholders? How are the public and private spheres shaped by ICTs? (Frissen 2000).

Civic discourse: How is the Internet utilized by women and women's groups for civic activities? How are nonexpert groups and citizens influencing the technological trajectory of society? How can technological culture be democratized? (Bijker and Bijsterveld 2000).

Design: Fountain asks: "how might the Information Society be affected if more women participate as experts and designers?" (2000, 55). Do women tech designers have a unique perspective? Would they, for instance, develop more social uses for applications, or engage technology users more in the initial design? In this case, I think that it's important to look at how young girls use the Internet, since many of them are intrepid Internet denizens.

Representation: How has the media industry addressed women as consumers and workers within the new economy of technological convergence through advertisements and specialized programming?

FINAL WORDS

> Cyberspace provides women with a new terrain on which to wage old struggles (Gallagher 1996).

Social engagement needs to occur at the collective level. We can use the Internet to debate, inform, organize, and mobilize, but at the end of the day, when we finally turn off the computer and peel our eyes from the monitor, we're still left with the everyday messiness of our lives. We still have bodies that need more than virtual connections, real children who need to eat real food and be reminded to do their homework, real dogs that need their daily park jaunts and real partners and friends that need conversation. No matter how much the tentacles of new media spread in their synergistic and adulterous couplings—and believe me, we ain't seen nothing yet!—we are all, at the end of the day, citizens. Not citizens.com.

REFERENCES

Adam, Alison. 2000. Feminist AI Projects and Cyberfutures. In *The gendered cyborg: A reader*, ed. Kirkup et .al. London: Routledge/The Open University: 276–290.

Adam, Alison. 1998. *Artificial knowing: Gender and the thinking machine*. New York: Routledge.

AMARC-WIN. The World Association of Community Radio Broadcasters-Women's International Network. *NGO alternative report-Beijing platform and community radio women,* presented to the Conference on Non-Governmental Organizations in Consultative Relationship with the United Nations (Congo), in Collaboration with the NGO Coordinating Committee for Beijing Plus Five. Montreal: April 14, 2000.

American Anthropological Society and Computing Research Association. 1996. *Culture, society and advanced information technology.* NSF Grant #CDA 9508431. URL: cra.org/Reports/Aspects

American Association of University Women (AAUW). 2000. *Tech-Savvy: Educating girls in the new computer age*. Washington, D. C.: AAUW.

Anderson, Lessley. January 28, 2000. It's Not Your Kid Sister's TigerBeat. *The Industry Standard*. URL: www.thestandard.com

Anderson, Robert H., et. al. 1995. *Universal access to e-mail: Feasibility and societal implications*. Santa Monica, CA: The Rand Corporation. URL: www.rand.org/publications/MR/MR650/

Arnold, June. Summer 1976. Feminist Presses and Feminist Politics. *Quest: A Feminist Quarterly* 3:18–26.

Association for Progressive Communication PC Women's Network, Gender and Information Technology Project. 1997. *Global networking for change: Experiences from the APC women's programme*.

Association for Progressive Communication, Women's Networking Support Program. *Information and Communication Technologies: A Women's Agenda*, 1999. URL: www.gn.apc.org/apcwomen/resources/policy/women-rights.html

Balka, Ellen. 1997a. Participatory design in women's organizations: The social world of organizational structure and the gendered nature of expertise. *Gender, Work and Organizations* 4:99–115.

Balka, Ellen. 1997b. *Computer networking: Spinsters on the web*. Ottawa: Canadian Research Institute for the Advancement of Women.

Balka, Ellen. July 1996. Women and computer networking in six countries. *The Journal of International Communication*: 66–84.

Balka, Ellen. 1993. Women's access to on-line discussions about feminism. *The Electronic Journal of Communication/La Revue Electronique de Communication* 3.

Balka, Ellen and Laurel Doucette. July 26, 1994. The accessibility of computers to organizations serving women in the province of Newfoundland: Preliminary study results. *Electronic Journal of Virtual Culture* 2.

Barber, Benjamin. 1995. *Jihad vs. McWorld*. New York: Times Books.

Barringer, Felicity. January 11, 2000. Does deal signal lessening of media independence? *The New York Times*. (Archived at www.nytimes.com)

Baym, Nancy. 2000. *Tune in, log on: Soaps, fandom, and online community*. Thousand Oaks, CA: Sage.

Beijing Declaration and Platform for Action. 1995. URL: women.usia.gov/usia/bei-jpg.htm.

Bell, Vickie and Denise de La Rue. 1995. *Gender harassment on the Internet*. Georgia State University College of Law. URL: www.gsu.edu/~lawppw/lawand.papers/harass.html

Benston, Margaret Lowe. 1989. Feminism and system design: Questions of control. In *The Effects of Feminist Approaches on Research Methodologies*. Waterloo, Ont.: Wilfrid Laurier University Press: 205–223

Berg, A. J. 1994a. Technological flexibility: Bringing gender into technology (or was it the other way around?). In *Bringing technology home: Gender and technology in a changing Europe*, ed. Cynthia Cockburn and Ruza Furst Dilic. Philadelphia: Open University Press: 94–110.

Berg, A. J. 1994b. A gendered socio-technical construction: The smart home. In *Bringing technology home: Gender and technology in a changing Europe*, ed. Cynthia Cockburn and Ruza Furst Dilic. Philadelphia: Open University Press: 65–180.

Biersdorfer, J. D. October 14, 1999. Trying on clothes in a virtual dressing room. *The New York Times* (Archived at www.nytimes.com).

Bijker, Wiebe E. and Karin Bijsterveld. July 2000. Women walking through plans: Technology, democracy, and gender identity. *Technology and Culture* 41:485–515.

Blanchfield, Mike. July 23, 2000. G-8 to strive for a wired world. *The Ottawa Citizen*: A6.

Bodker, Suzanne and Joan Greenbaum. 1993. Design of information systems: Things versus people. In *Gendered by design: Information technology and office systems*, ed. Green, Owen, and Pain. Washington: Taylor and Francis: 53–63.

Borowy, Jan and Theresa Johnson. 1995. Unions confront work reorganization and the rise of precarious employment: Home-based work in the garment industry and federal public service. In *Re-shaping work: Union responses to technological change*, ed. C. Schenk and J. Anderson. Don Mills, ON: Ontario Federation of Labour, Technology Adjustment Research Programme: 29–47.

Borsook, Paulina. 2000. *Cyberselfish: A critical romp through the terribly libertarian culture of high-tech*. New York: PublicAffairs.

Brail, Stephanie. 1996. The price of admission: Harassment and free speech in the wild, wild west. In *Wired women: Gender and new realities in cyberspace*, ed. Cherny and Weise. Seattle, WA: Seal Press: 141–157.

Brier, Steven E. March 29, 2000. Since women ask for directions, the web is being revamped. *The New York Times*. (Archived at www.nytimes.com).

Brown, Janelle. April 4, 1999. Beauty and the geeks. *The Ottawa Citizen*: C4. Originally published in *Salon Magazine* (www.salon.com).

Brownmiller, Susan. 1999. *In our time: Memoir of a revolution*. New York: The Dial Press.

Brunsdon, Charlotte. 2000. *The feminist, the housewife, and the soap opera*. Oxford: Oxford University Press.

Butsch, Richard. October 1998. Crystal sets and scarf-pin radios: Technology and the construction of American radio listening in the 1920s. *Media, Culture & Society* 20:556.

Carstensen, Jeanne. Spring 2000. Relinquishing the MIC. *Whole Earth Review*: 90–92.

Cassell, Justine and Henry Jenkins, eds. 1998. *From Barbie to Mortal Kombat: Gender and computer games*. Cambridge, MA: The MIT Press.

Castells, Manuel. 1997. The power of identity. *The information age: Economy, society and culture*, vol 2. Oxford: Blackwell Publishers.

Cauley, Leslie. July 24, 1995. Phone giants discover the interactive path is full of obstacles. *The Wall Street Journal*: A1.

Center for Media Education. January 10, 2000. Consumer groups respond to AOL-Time Warner deal. URL: www.cme.org/press/000110pr.html.

Cherny, Lynn and Elizabeth Reba Weise. 1996. *wired_women: Gender and new realities in cyberspace*. Seattle, WA: Seal Press.

Church, Elizabeth. July 20, 2000. Travel wanderlust leads journeywomen to the web. *The Globe and Mail*: B10.

Clark, Tim. December 31, 1998. Women shoppers crowd internet stores. *CNET News.com*. URL: news.cnet.com/category/0-1007-200-336893.html

Cleaver, Harry M., Jr. Spring 1998. The Zapatista effect: The internet and the rise of an alternative political fabric. *Journal of International Affairs* 51: 621.

Clement, Andrew and Leslie Regan Shade. 2000. The access rainbow: Conceptualizing universal access to the information/communications infrastructure. In *Community informatics: Enabling communities with information and communication technologies*, ed. Michael Gurstein. Hershey, PA: Idea Group Publishing: 32–51.

Cockburn, Cynthia and Susan Ormrod. 1993. *Gender and technology in the making*. Thousand Oaks: Sage.

Cole, Elizabeth R. and Andrea Lee Press. 1999. *Speaking of abortion: Television and authority in the lives of women*. Chicago: University of Chicago Press.

Collins-Jarvis, Lori. 1993. Gender representation in an electronic city hall: Female adoption of Santa Monica's PEN system. *Journal of Broadcasting and Electronic Media* 27: 49–66.

Cooper, Cameron. November 22, 1999. Japan's cyber women to win job war. *Business Asia* 7:1.

Covert, Catherine L. 1992. 'We may hear too much': American sensibility and the response to radio, 1919–1924. In *Media voices: Our historical perspective*, ed. Jean Folkerts. New York: Macmillan Pub. Co.: 300–315.

Cowan, Ruth Schwartz. 1989. The consumption junction: A proposal for research strategies in the sociology of technology. In *The social construction of technological systems: New directions in the sociology and history of technology*, ed. Bijker, Hughes, and Pinch. Cambridge, MA: The MIT Press: 261–280.

Cowan, Ruth Schwartz. 1983. *More work for mother: The ironies of household technology from the open hearth to the microwave*. New York: Basic Books.

Coyle, Karen. 1996. How hard can it be? In *wired_women: Gender and new realities in cyberspace*, ed. Cherny and Weise. Seattle, WA: Seal Press: 42–55.

Credé, Andreas and Robin Mansell. 1998. *Knowledge societies…in a nutshell: Information technology for sustainable development*. Ottawa: IDRC.

Cryderman, Kelly. August 8, 2000. Cross-stitchers keep an eye on the net. *The Ottawa Citizen*: A4.

Cuklanz, Lisa M. 1999. *Rape on prime time: Television, masculinity, and sexual violence*. Philadelphia: University of Pennsylvania Press.

Currie, Dawn H. 1999. *Girl talk: Adolescent magazines and their readers*. Toronto: University of Toronto Press.

Dam, Kenneth and H. Lin, eds. 1996. *Cryptography's role in securing the information society*. Committee to Study National Cryptography Policy, Computer Science and Telecommunications Board, Commission on Physical Science, Mathematics and Applications, National Research Council. Washington: National Academy Press. [May 30, 1996, prepublication copy]

Davis, Dineh M. 1998. Women on the net: Implications of informal international networking surrounding the Fourth World Conference on Women. In *Digital democracy: Policy and politics in the wired world*, ed. Cynthia J. Alexander and Leslie A. Pal. Toronto: Oxford University Press, 87–104.

Dean, Jodi. 1999. Virtual fears. *Signs: Journal of Women in Culture and Society* 25:1069–1078.

Demont, Philip. September 16, 1999. Hollinger, drug chain create handbag.com for women. *The Ottawa Citizen*: C1.

Dery, Mark. 1996. *Escape velocity: Cyberculture at the end of the century*. New York: Grove Press.

de Zegher, Catherine, ed. 1998. *Martha Rosler: Positions in the life world*. Birmingham: Ikon Gallery; Vienna: Generali Foundation; Cambridge, MA: The MIT Press.

Douglas, Susan J. 1999. *Listening in: Radio and the American imagination*. New York: Times Books.

Douglas, Susan J. 1987. *Inventing American broadcasting, 1899–1922*. Baltimore, MD: The Johns Hopkins University Press.

Ebden, Theresa. June 15, 2000. More Canadian women shopping online. *The Globe and Mail*: B9.

Eisenstein, Zillah. 1998. *Global obscenities: Patriarchy, capitalism, and the lure of cyberfantasy*. New York: New York University Press.

Escobar, Arturo. 1999. Gender, place and networks: A political ecology of cyberculture. In *Women@Internet: Creating new cultures in cyberspace*, ed. Wendy Harcourt. New York: Zed Books, 31–54.

Everts, Saskia. 1998. *Gender and technology: Empowering women, engendering development*. London; New York: Zed Books.

Farwell, Edie, Peregrine Wood, Maureen James, and Karen Banks. 1999. Global networking for change: Experiences from the APC women's programme. In *Women@Internet: Creating new cultures in cyberspace*, ed. Wendy Harcourt. New York: Zed Books, 1999.

Fischer, Claude S. 1992. *America calling: A social history of the telephone to 1940*. Berkeley: University of California Press.

Flowers, Amy. 1998. *The fantasy factory: An insider's view of the phone sex industry*. Philadelphia: University of Pennsylvania Press.

Flynn, Laurie J. February 8, 1999. Microsoft is starting web site aimed at big audience: women. *The New York Times*. (Archived at www.nytimes.com).

Fountain, Jane E. 2000. Constructing the information society: Women, information technology, and design. *Technology in Society* 22:45–62.

Franklin, Ursula. 1990. *The real world of technology*. Toronto: CBC Enterprises.

Frankson, Joan Ross. July 1996. Women's global faxnet: Charting the way. *The Journal of International Communication* 3(1):102–110.

Frederick, Howard. 1992. Computer communications in cross-border coalition-building: North American NGO networking against NAFTA. *Gazette: The International Journal for Mass Communication Studies* 50:217–241.

Freeman, Alexa and Valle Jones. Fall 1976. Creating feminist communications. *Quest: A Feminist Quarterly* 3: 3–10.

Freeman, Carla. 2000. *High tech and high heels in the global economy: Women, work, and pink-collar identities in the Caribbean*. Durham: Duke University Press.

Frissen, Valerie A. J. 2000. ICTS in the rush hour of life. *The Information Society*: 65–75.

Froehling, Oliver. 1999. Internauts and guerrilleros: The Zapatista rebellion in Chiapas, Mexico and its extension into cyberspace. In *Virtual geographies: Bodies, space and relations*, ed. Mike Crang, Phil Crang and Jon May. New York: Routledge, 164–177.

Gadd, Jane. August 12, 1995. Where women stand now. *The Globe and Mail*: A5.

Gajjala, Radhika and Annapurna Mamidipudi. 1999. Cyberfeminism, technology and international development. In *Gender and technology*, ed. Caroline Sweetman. Oxford: Oxfam, 8–16.

Galegher, Jolene, Lee Sproull, and Sara Kiesler. October 1998. Legitimacy, authority, and community in electronic support groups. *Written Communication* 15:493.

Gallagher, Margaret. 1996. Lipstick imperialism and the new world order: Women and media at the close of the twentieth century. Paper prepared for Division for the Advancement of Women, Department for Policy Coordination and Sustainable Development, United Nations. URL: www.un.org/esa/documents/esc/cn6/1996/media/Gallagh.htm

Gallagher, Margaret and Lilia Quindoza-Santiago, eds. 1994. *Women empowering communication: A resource book on women and the globalisation of media.* London, Manila, New York: World Association for Christian Communication, Isis International, International Women's Tribune Centre.

Garrubbo, Gina. February 2, 1998. What makes women click? *Advertising Age* 69(5):12.

Gittler, Alice Mastrangelo. 1999. Mapping women's global communications and networking. In *Women@Internet: Creating new cultures in cyberspace*, ed. Wendy Harcourt. London: Zed Books, 91–101.

Gittler, Alice Mastrangelo. July 1996. Taking hold of electronic communications: Women making a difference. *The Journal of International Communication* 3:85–101.

Glusker, Susannah. October 1998. Women networking for peace and survival in Chiapas: Militants, celebrities, academics, survivors, and the stiletto heel brigade. *Sex Roles: A Journal of Research* 39:539.

Goldberg, Rosalee. 1998. *Performance art: From futurism to the present.* New York.: Henry Abrams.

Graphic evidence: Web grrls. August 12, 2000. *The Globe and Mail*: A8.

Green, Eileen and Alison Adam. 1999. On-line leisure: Gender and ICTs in the home. *Information, Communication and Society* 1:291–312.

Green, Eileen, Jenny Owen, and Den Pain, eds. 1993. *Gendered by design?: nformation technology and office systems.* Washington: Taylor & Francis.

Greenbaum, Joan. June 1990. The head and the heart: Using gender analysis to study the social construction of computer systems. *Computers and Society* 20:9–17.

Grusin, Richard and Jay David Bolter. 1999. *Remediation: Understanding new media.* Cambridge, MA: The MIT Press.

Guillermo, Delgado-P. and Marc Becker. Winter 1998. Latin America: The Internet and indigenous texts. *Cultural Survival Quarterly* 21. URL: www.cs.org/CSQ/csqinternet.html#Becker

Gunnarsson, Ewa and Ursula Huws. 1997. *Virtually free: Gender, work and spatial choice.* Stockholm: Swedish National Board for Industrial and Technical Development.

Gurstein, Michael, ed. 2000. *Community informatics: Enabling communities with information and communication technologies.* Hershey, PA: Idea Group Publishing.

Hacker, Sally. 1979. Sex stratification and organizational change: A longitudinal study of AT&T. *Social Problems* 26:539–557.

Hamelink, Cees J. 1999. Human development. In *World communication and information report 1999–2000.* UNESCO. URL: www.unesco.org/webworld/wcir/en/report.html

Hamm-Greenawalt, Lisa. March 15, 2000. Babes in boyland. *Internet World* 6:71.

Hansell, Saul. January 11, 2000. America Online to buy Time Warner for $165 billion". *The New York Times.* (Archived at www.nytimes.com).

Haralovich, Mary Beth and Lauren Rabinovitz.1999. Introduction to *Television, history, and American culture: Feminist critical essays*, ed. Haralovich and Rabinovitz. Durham: Duke University Press, 1–16.

Harris, Roma. 2000. Squeezing librarians out of the middle. In *Women, work and computerization: Charting a course to the future*, ed. Ellen Balka and Richard Smith. Boston: Kluwer Academic Publishers, 250–259.

Hawthorne, Susan and Renate Klein. 1999. Introduction to *CyberFeminism: Connectivity, critique, creativity*, ed. Hawthorne and Klein. Melbourne: Spinifex Press, 1–19.

Headlam, Bruce. January 20, 2000. Barbie PC: Fashion over logic. *The New York Times*: E4.

Herman, Edward S. and Robert W. McChesney. 1997. *The global media: The new missionaries of global capitalism*. Washington: Cassell.

Herz, J.C. 1995. *Surfing on the Internet: A nethead's adventures on-line*. Boston: Little, Brown and Company.

Hilmes, Michele. 1997. *Radio voices: American broadcasting, 1922–1952*. Minneapolis: University of Minnesota Press.

Hopkins, Patrick D., ed. 1998. *Sex/machine: Readings in culture, gender, and technology*. Bloomington, Indiana: Indiana University Press

Hotwired. *Encyclopedia of the new economy*. URL: www.hotwired.com/special/ene/

Hu, Jim. January 21, 2000. Study: Net's gender gap narrows. *CNET news.com*. URL: news.cnet.com/news/0-1005-200-1528683.html

Huws, Ursula. 1999. Women, participation and democracy in the information society. Paper written for *Citizens at the Crossroads* conference, University of Western Ontario, London, Ontario. URL: dialspace.dial.pipex.com/town/parade/hg54/xroad.htm

Huyer, Sophia. 1999. Shifting agendas at GK97: Women and international policy on information and communication technologies. In *Women@Internet: Creating new cultures in cyberspace*, ed. Wendy Harcourt. London: Zed Books, 114–130.

I am cyber-women, hear me roar. November 15, 1999. *Business Week*: 40.

IDRC Gender and Information Working Group. 1995. Information as a trans-formative tool. In *Missing links: Gender equity in science and technology for development*. Gender Working Group, United Nations Commission on Science and Technology for Development. Ottawa: International Development Research Centre: 267–293.

Industry Canada. December 3, 1997. Canada and the knowledge-based economy. IT and Knowledge Based Economy Summit. URL: http://strategis.ic.gc.ca

Information Highway Advisory Council. 1997. *Preparing Canada for a digital world*. Ottawa: Industry Canada. URL: strategis.ic.gc.ca/ssg/ih01650e.html.

Inness, Sherrie A., ed. 1998a. *Delinquents and debutantes: Twentieth-century girls' cultures*. New York: New York University Press.

Inness, Sherrie A., ed. 1998b. *Millennium girls: Today's girls around the world*. Lanham, MD: Rowman and Littlefield Publishers, Inc.

International Telecommunication Union (ITU). 1999. *Challenges to the network: Internet for development*. Geneva: ITU. URL: www.itu.org

International Telecommunication Union (ITU). 1998. *World telecommunication development report 1998 — Universal access*. Geneva: ITU. URL: www.itu.org

Internet clampdown in Mecca. April 17, 2000. *BBC News Online*. URL: news.bbc.co.uk

Ippolito, Jon. July–August 2000. Gender changer. *Artbyte: The Magazine of Digital Culture*: 36–39.

Jeffrey, Liss. 1995. The doors of reception. In *Watching TV: Historic televisions and memorabilia from the MZTV Museum*. Toronto: Royal Ontario Museum, 10–25.

Johannson, Ingrid. January 1998. Women on the Internet: How Swede it is! *Direct Marketing* 60:26.

Johnson, Greg. January 31, 2000. Women closing fast on men's Internet lead, as advertisers note rising interest. *Providence Business News* 14:12B.

Johnson, Lesley. 1988. *The unseen voice: A cultural history of early Australian radio.* London: Routledge.

Jones, Amelia. 1998. *Body art: Performing the subject.* Minneapolis: University of Minnesota Press.

Juffer, Jane. 1998. *At home with pornography: Women, sex, and everyday life.* New York: New York University Press.

Kalakota, Ravi and Andrew B. Whinston. 1996. *Frontiers of electronic commerce.* Reading, MA: Addison-Wesley.

Kaminer, Wendy. March 9, 1998. The chador hits cyberspace: Protection of female modesty always leads to inequality, even online. *The Nation* 266:21.

Kaufman, Leslie. March 9, 2000. The dot-com world opens new opportunities for women to lead. *The New York Times.* (Archived at www.nytimes.com).

Kearney, Mary Celeste. 1998. Producing girls: Rethinking the study of female youth culture. In *Delinquents and debutantes: Twentieth-century girls' cultures,* ed. Sherrie A. Inness. New York: New York University Press, 285–310.

Kendall, Lori. April 2000. 'OH NO! I'M A NERD!': Hegemonic masculinity in an online forum. *Gender and Society* 14:256–274.

Kerwin, Ann Marie. June 1999. 'GH,' Women.com roll holiday shopping guide — Reliability test a requirement for advertisers. *Advertising Age* (Archived at www.adage.com).

Kilbourne, Jean. 1999. *Deadly persuasions: Why women and girls must fight the addictive power of advertising.* New York: The Free Press.

Kirkup, Gill, Linda Janes, Kath Woodward, and Fiona Hovenden. 2000. *The gendered cyborg: A reader.* London: Routledge/The Open University.

Klawe, Maria and Nancy Leveson. January 1995. Women in computing: Where are we now? *Communications of the ACM* 38:29–35.

Klein, Naomi. 2000. *No logo: Taking aim at the brand bullies*. Toronto: Knopf Canada.

Kling, Rob, Mark S. Ackerman, and Jonathan P. Allen. 1995. Information entre-preneuralism, information technologies, and the continuing vulnerability of privacy. In *Computerization and controversy: Value conflicts and social choices*, 2nd ed. Rob Kling, ed. San Diego: Academic Press.

Kole, Ellen S. August 1998. Myths and realities in Internet discourse: Using computer networks for data collection and the Beijing World Conference on Women. *Gazette: The International Journal for Mass Communication Studies* 60:343–360.

Koranteng, Juliana. December 1999. Sexes at variance in use of Internet. *Advertising Age International*: 21.

Korenman, Joan. 1999. Email forums and women's studies: The example of WMST-L. In *CyberFeminism: Connectivity, critique, creativity*, ed. Susan Hawthorne and Renate Klein. Melbourne: Spinifex Press, 80–97.

Kramarae, Cheris. 1997. Technology policy, gender, and cyberspace. *Duke Journal of Gender Law and Policy* 149.

Kramer, Jana and Cheris Kramarae. 2000. Women's political webs: Global elec-tronic networks. In *Gender, politics and communication*, ed. Annabelle Sreberny and Liesbet van Zoonen. Cresskill, NJ: Hampton Press, 205–222.

Kuchinskas, Susan. September 7, 1998. It's a women's web. *Brandweek*:46.

Larson, Erik. October 11, 1999. Free money. *The New Yorker*: 76–85.

Laucius, Joanne. August 10, 2000. Girls are taking over the web. *The Ottawa Citizen*: A1.

Ledbetter, James. September 18, 1998. Men are from mars, women are from AOL. *The Industry Standard*. URL: www.thestandard.com/article/display/0,1151,1750,00.html

Lee, Samantha. May 29, 2000. What do women want? *Forbes*: 42.

Lee, Sarah. December 1999. Private uses in public spaces: A study of an Internet cafe. *New Media and Society* 1:331–350.

Li, Kenneth. December 16, 1999. Oxygen Media flashes the cash. *The Industry Standard*. URL: www.thestandard.com/article/display/0,1151,8244,00.html

Light, Jennifer. 1995. The digital landscape: New space for women? *Gender, Place and Culture* 2:133–146.

Liming, Wei. August 10, 1998. Women and the Internet. *Beijing Review* 41:37.

Lohan, Maria. 2000. Come back public/private; (almost) all is forgiven: Using feminist methodologies in researching information communication technologies. *Women's Studies International Forum* 23:107–117.

Macavinta, Courtney. November 5, 1999. Women's sites seek to separate from the crowd. *CNET News.com*. Archived at news.cnet.com

MacKenzie, Donald and Judy Wajcman, eds. 1999. *The social shaping of technology*, 2nd ed. Buckingham: Open University Press.

MacKinnon, Catharine A. June 1995. Vindication and resistance: A response to the Carnegie Mellon study of pornography in cyberspace. *The Georgetown Law Journal* 83.

Maddox, Brenda. 1977. Women and the switchboard. In *The social impact of the telephone*, ed. Ithiel de Sola Pool. Cambridge, MA: MIT Press, 262–280.

Madon, Shirin. 2000. The Internet and socio-economic development: Exploring the interaction. *Information Technology and People* 13:85–102.

Manegold, Catherine S. July 12, 1992. No more nice girls. *The New York Times*: 25.

Mansell, Robin and Ute Wehn. 1998. *Knowledge societies: Information technology for sustainable development*. Oxford: Oxford University Press.

Mara, Janis. March 20, 2000. Women.com and E! TV ink cross-promo pact. *Brandweek* 42:76.

Mara, Janis. May 22, 2000. The halo effect. *Brandweek* 41:86.

Marcelle, Gillian M. 2000. Getting gender into African ICT policy: A strategic view. In *Gender and the Information Revolution in Africa*, ed. Eva M. Rathgeber and Edith Ofwona Adera. Ottawa: International Development Research Centre, 35–83.

Marcelle, Gillian M. 1997. *Using information technology to strengthen African women's organizations*. London: Abantu Publications.

Markle Foundation/Oxygen Media, Inc. 2000. Research and experimentation for converging media. URL: www.markle.org/programs/pit/Project.200003071416.1621.html

Martin, Michele. 2001. An un-suitable technology for a woman?: Communication as circulation. In *Sex and money: Feminism and political economy in the media*, ed. Eileen R. Meehan and Ellen Riordan. Minneapolis: University of Minnesota Press.

Martin, Michele. 1991. *'Hello Central?': Gender, technology, and culture in the formation of telephone systems*. Montreal: McGill-Queen's University Press.

Martin, Scott. December 15, 1998. More women shopping online. *CNET News.com*. URL: news.cnet.com/news/0-1007-200-336462.html

Marvin, Carolyn. 1987. *When old technologies were new: Thinking about electric communication in the late nineteenth century*. New York: Oxford University Press.

Mathews, Jessica T. January/February 1997. Power shift. *Foreign Affairs:* 50–66.

McChesney, Robert. 2000. So much for the magic of technology and the free market: The World Wide Web and the corporate media system. In *The World Wide Web and contemporary cultural theory*, ed. Andrew Herman and Thomas Swiss. New York: Routledge, 5–35.

McChesney, Robert W. 1999. *Rich media, poor democracy: Communication politics in dubious times.* Urbana: University of Illinois Press.

McLaughlin, Lisa. October 1999. Beyond 'separate spheres': Feminism and the cultural studies/political economy debate. *Journal of Communication Inquiry* 23:327–354.

McQuiston, Liz. 1997. *Suffragettes to she-devils: Women's liberation and beyond.* London: Phaidon Press Ltd.

Meehan, Eileen R. 2000. Feminism and political economy: On not being a hot commodity. The Dallas Smythe Memorial Lecture, Simon Fraser University, Vancouver, B.C., Canada, April 14.

Meehan, Eileen R. 1990. Why We Don't Count. In *Logics of television*, Patricia Mellencamp, ed. Bloomington: Indiana University Press, 117–137.

Meehan, Eileen R. 1984. Ratings and the institutional approach. *Critical Studies in Mass Communication* 1:216–225.

Meehan, Eileen and Mia Consalvo. October 1999. Introducing the issues — An interview with Eileen Meehan. *Journal of Communication Inquiry* 23: 321–326.

Meehan, Eileen and Ellen Riordan, eds. 2001. *Sex and money: Feminism and political economy in the media.* Minneapolis: University of Minnesota Press.

Menzies, Heather. Summer 1997. Telework, shadow work: The privatization of work in the new digital economy. *Studies in Political Economy* 53:103–143.

Miller, Francesca. 1991. *Latin American women and the search for social justice.* Hanover, NH: University Press of New England.

Miller, Heather Lee. Autumn 1999. Getting to the source: The World Wide Web of resources for women's history. *Journal of Women's History* 11:176.

Miller, Laura. 1995. Women and children first: Gender and the settling of the electronic frontier. In *Resisting the virtual life: The culture and politics of information*, ed. James Brook and Iain A. Boal. San Francisco: City Lights, 49–57.

Missing links: Gender equity in science and technology for development. 1995. Gender Working Group, United Nations Commission on Science and Technology for Development. Ottawa: International Development Research Centre.

Mitter, Swasti. 2000. Women in knowledge societies. Keynote address to the Global Knowledge II Conference, Kuala Lumpur, Malaysia, March. Available at URL: www.globalknowledge.org.my/

Mitter, Swasti and Sheila Rowbotham, eds. 1997. *Women encounter technology: Changing patterns of employment in the third world.* New York: Routledge.

Mittner, Greta. January 29, 1999. *Hearst gives muscle to women.* Red Herring Online. URL: www.redherring.com.

Mosco, Vincent. 2000. Webs of myth and power: Connectivity and the new computer technopolis. In *The World Wide Web and contemporary cultural theory*, ed. Andrew Herman and Thomas Swift. New York: Routledge, 37–60.

Mosco, Vincent. 1996. *The political economy of communication.* Thousand Oaks, CA: Sage.

Moyal, Ann. 1992. The gendered use of the telephone: An Australian case study. *Media, Culture and Society* 14:51–72.

Moyal, Ann. 1989. *Women and the telephone in Australia.* Study prepared for Telecom Australia, Strategic Analysis Unit, Corporate Directorate, Telecom Australia, Melbourne.

Nussbaum, Martha C. 2000. *Women and human development: The capabilities approach.* London: Cambridge University Press.

Oldfield, Margaret. May/June 1995. Heaven or hell: Telework & self-employment. *Our Times*:16–119.

Oldfield, Margaret. July 1991. The electronic cottage—boon or bane for mothers?: Observations from an empirical study of mothers doing clerical work on computers at home. Paper presented to the Fourth International Conference on Women, Work and Computerization, Helsinki, Finland.

Olson, Stephanie. October 15, 1999. *Beauty sites make over the web.* CNETnews.com. URL: www.news.cnet.com/news/0-1007-200-853532.html.

Ong, Aihwa. 1987. Disassembling gender in the electronics age. *Feminist Studies* 13:609–626.

Online in Saudi Arabia: How women beat the rules. October 2, 1998. *The Economist* 353:48.

On the eve of the future: A secret history of women and technology. July–August 2000. *Artbyte: The Magazine of Digital Culture*: 26–30.

Opoku-Mensah, Aida. 2000. ICTs as tools of democratization: African women speak out. In *Gender and the information revolution in Africa*, ed. Eva M. Rathgeber and Edith Ofwona Adera. Ottawa: International Development Research Centre, 187–213.

Oppenheimer, Andres. 1998. *Bordering on chaos: Mexico's roller-coaster journey toward prosperity.* Boston: Little, Brown and Company.

Organization for Economic Cooperation and Development (OECD). 1997. *Electronic commerce: Opportunities and challenges for government* (The "Sacher Report"). URL: www.oecd.org/dsti/sti/it/ec/act/sacher.htm

Organization for Economic Cooperation and Development (OECD). 1996. *The knowledge-based economy.* Paris: OECD. URL: www.oecd.org/dsti/sti/s_t/inte/prod/kbe.htm

Oxygen media announces online advertising agreement with MSN Hotmail. January 17, 2000. MSN Press Release. URL: www.microsoft.com/press-pass/press/2000/Jan00/Oxygenpr.asp

Panos Institute. 1996. *The Internet & the South: Superhighway or Dirt-Track?* URL: www.netural.com/lip/file_a03.html

Partridge, John. May 25, 2000. Toronto entrepreneur locks horns with Oprah. *The Globe and Mail*: B4.

Peer, Limor. 2000.Women, talk radio, and the public sphere(s) in the United States. In *Gender, politics, and communication*, ed. Annabelle Sreberny and Liesbet van Zoonen. Cresskill, NJ: Hampton Press, 299–327.

Pew Internet & American Life Project. May 10, 2000. *Tracking online life: How women use the Internet to cultivate relationships with family and friends*. URL: www.pewinternet.org

P&G targets women with new beauty products site. September 13, 1999. Reuters — Special to *CNET News.com*. URL: news.cnet.com/news/0-1007-200-117s560.html

Plant, Sadie. 2000. On the matrix: Cyberfeminist simulations. In *The gendered cyborg*, ed. Gill Kirkup, Linda Janes, Kath Woodward, and Fiona Hovenden. London: Routledge/The Open University, 265–275. Originally published in *Cultures of the Internet: Virtual spaces, real histories, living bodies*. 1996. ed. Rob Shields. Thousand Oaks, CA: Sage.

Plant, Sadie. 1997. *Zeros+ones: Digital women + the new technoculture*. New York: Doubleday.

Pollock, Scarlett and Jo Sutton. 1998. Women click: Feminism and the Internet. In *CyberFeminism: Connectivity, critique and creativity*, ed. Susan Hawthorne and Renate Klein. Melbourne: Spinifex Press, 33–50.

Pollock, Scarlett and Jo Sutton, eds. 1997. *Virtual organizing, real change: Women's groups using the Internet*. Ottawa: Women'space.

Prose, Francine. February 13, 2000. A wasteland of one's own. *The New York Times Magazine*: 66–71.

Raik-Allen, Georgie. June 19, 1999. Paul Allen invests $100 million in wired women. *Redherring.com*. URL: www.redherring.com/insider/1999/0619/vc-oxygen.html

Rakow, Lana F. 1992. *Gender on the line: Women, the telephone, and community life*. Chicago: University of Illinois Press.

Rakow, Lana F. 1988a. Gendered technology, gendered practice. *Critical Studies in Mass Communication* 5:57–70.

Rakow, Lana. 1988b. Women and the telephone: The gendering of a communications technology. In *Technology and women's voices: Keeping in touch*, ed. Cheris Kramarae. New York: Routledge & Kegan Paul, 207–228.

Rakow, Lana F. Autumn 1986. Rethinking gender research in communication. *Journal of Communication* 36:11–26.

Rakow, Lana and Vija Navarro. Winter 1993. Remote mothering and the parallel shift: Women meet the cellular telephone. *Critical Studies in Mass Communication* 10:144–157.

Rapping, Elayne. July 2000. You've come which way, baby? *The Women's Review of Books* Vol. 17:20–22.

Rathgeber, Eva M. and Edith Ofwona Adera, eds. 2000. *Gender and the information revolution in Africa*. Ottawa: International Development Research Centre.

Reddick, Andrew. 2000. *The dual digital divide: The information highway in Canada*. Ottawa: Public Interest Advisory Centre. URL: olt-bta.hrdc-drhc.gc.ca/download/oltdualdivideen.pdf

Resnick, Rosalind. 1995. Survey of women online. *Interactive Publishing Alert*. URL: www.netcreations.com/ipa/women.htm

Riano, Pilar, ed. 1994. *Women in grassroots communication: Furthering social change.* Thousand Oaks, CA: Sage.

Rius, Marisa Belausteguigoitia. 1999. Crossing borders: From crystal slippers to tennis shoes. In *Women@Internet: Creating new cultures in cyberspace,* ed. Wendy Harcourt. London: Zed Books, 23–30.

Romero, Simon. March 28, 2000. Weavers go dot-com, and elders move in. *The New York Times* (Archived at www.nytimes.com).

Rommes, Els, Ellen van Oost, and Nelly Oudshoorn. Winter 1999. Gender in the design of the virtual city of Amsterdam. *Information, Communication and Society* 2(4):476–495.

Ronfeldt, David, John Arquilla, Graham E. Fuller, and Melissa Fuller. 1998. *The Zapatista social netwar in Mexico.* Santa Monica, CA: The Rand Corporation.

Rosenberg, Jessica and Gitana Garofalo. 1998. Riot grrrl: Revolutions from within. *Signs: Journal of Women in Culture and Society* 23(3):809–841.

Sandoval, Greg. May 26, 2000. Women top shoppers online, study finds. *CNET News.com.* URL: news.cnet.com/news/0-1007-200-1956977.html

Sandoval, Greg. April 10, 2000. CondeNet to launch women's fashion site. *CNET News.com.* (Archived at news.cnet.com).

Sassen, Saskia. 1998. Notes on the incorporation of third world women into wage labor through immigration and offshore production. In *Globalization and its discontents: Essays on the new mobility of people and money.* New York: The New Press, 81–109.

Scheyvens, Regina and Helen Leslie. 2000. Gender, ethics and empowerment: Dilemmas of development fieldwork. *Women's Studies International Forum* 23(1):119–130.

Schiller, Dan. 1999. *Digital capitalism: Networking the global market system.* Cambridge, MA: The MIT Press.

Schor, Juliet B. 1992. *The overworked American*. New York: Basic Books.

Schreiner, Heather. 1999. Rural women, development, and telecommunications: A pilot programme in South Africa. In *Gender and technology*, ed. Caroline Sweetman. Oxford: Oxfam, 64–70.

Schuler, Doug. 1996. *New community networks: Wired for change*. New York: ACM Press/Addison-Wesley Publishing Company.

Schuler, Doug and Aki Namioka. 1993. *Participatory design: Principles and practices*. Hillsdale, NJ: Lawrence Erlbaum Associates, Publishers.

Scodari, Christine. Fall 1998. 'No politics here': Age and gender in soap opera 'cyberfandom'. *Women's Studies in Communication* 21:168.

Sefton, Dru. March 16, 2000. Teen girls feel the net effect. *USA Today*: 3D.

Sennett, Richard. 1999. *The corrosion of character: The personal consequences of work in the new capitalism*. New York: W.W. Norton.

Shade, Leslie Regan. 1999. Net gains: Does access equal equity? *Journal of Information Technology Impact* 1:23–39. URL: www.jiti.com

Shade, Leslie Regan. 1997a. *Public space*. A backgrounder paper for the workshop, Developing a Canadian Access Strategy: Universal Access to Essential Network Services, February 6–8, 1997. Toronto: Information Policy Research Project. URL: www.fis.utoronto.ca/research/iprp/ua/ps.html#5

Shade, Leslie Regan. 1997b. *Using a gender-based analysis in developing a Canadian access strategy: Backgrounder report*. Prepared for the Ad Hoc Committee for the Workshop on Access to the Information Highway, Faculty of Information Studies, University of Toronto. URL: www.fis.utoronto.ca/research/iprp/ua/gender/GenderBased.html

Shade, Leslie Regan. 1997c. *Gender and community in the social constitution of the Internet*. Ph.D.diss., Graduate Program in Communications, McGill University, Montreal.

Shade, Leslie Regan. Winter 1996. Women, the World Wide Web, and issues of privacy. *Feminist Collections* 17:33–35. URL: www.library.wisc.edu/libraries/WomensStudies/fcshade.htm

Shade, Leslie Regan. 1994. Gender issues in computer networking. In *Women, work and computerization: Breaking old boundaries, building new forms,* ed. Adam, et. al. Amsterdam: Elsevier, 91–105.

Sharf, Barbara F. Winter 1997. Communicating breast cancer on-line: Support and empowerment on the Internet. *Women and Health* 2:65.

Shiva, Vandana. 1997. *Biopiracy: The plunder of nature and knowledge.* Toronto: Between the Lines.

Shneiderman, Ben. January 1995. The information superhighway: For the people. *Communications of the ACM* 38:162.

Shneiderman, Ben. May 2000. Universal usability. *Communications of the ACM* 43:85–91.

Silliman, Jael. 1999. Expanding civil society, shrinking political spaces: The case of women's nongovernmental organizations. In *Dangerous intersections: Feminist perspectives on population, environment, and development.* Cambridge, MA: South End Press, 133–162.

Singh, Supriya. 1998. *Electronic commerce and the gender imbalance.* Working Paper 1998/3, CIRCIT at RMIT (The Centre for International Research on Communication and Information Technologies, RMIT University, Australia). URL: www.circit.rmit.edu.au/publics/wp19983.html

Skrzeszewski, Stan and Maureen Cubberley. March 28, 1995a. *Canada's public libraries and the information highway: A report prepared for Industry Canada.* Toronto: The Coalition for Public Information/The Ontario Library Association.

Skrzeszewski, Stan and Maureen Cubberley. 1995b. *Future knowledge: The report.* Toronto: The Coalition for Public Information/The Ontario Library Association. URL: www.nlc-bnc.ca/documents/infopol/canada/cpi-fk.txt

Skrzeszewski, Stan and Maureen Cubberley. December 6, 1995c. A Response from Canada's Coalition for Public Information to Connection, Community, Content. Toronto: CPI. URL: www.canarie.ca/cpi.

Smith, J. and Ellen Balka. 1988. Chatting on feminist computer networks. In *Technology and women's communication*, ed. Cheris Kramarae. London: Routledge & Kegan Paul, 82–97.

Sofia, Zoe. 1998. The mythic machine: Gendered irrationalities and computer culture. In *Education/technology/power: Educational computing as a social practice*, ed. Hank Bromley and Michael W. Apple. Albany: State University of New York Press, 29–51.

Spender, Dale. 1995. *Nattering on the net: Women, power and cyberspace*. Melbourne: Spinifex Press.

Spertus, Ellen and Evelyn Pine. Winter 2000. Gender in the Internet age. *CPSR Newsletter* 18. URL: www.cpsr.org/publications/newsletters/issues/2000/winter2000

Spigel, Lynn. 1992. *Make room for TV: Television and the family ideal in postwar America*. Chicago: The University of Chicago Press.

Spiro, Leah Nathans. February 2000. In search of glamazon. *Talk*: 65–67.

Spreading the word about breast cancer. Winter 1999. *Human Ecology Forum* 27:21.

Sreberny, Annabelle. 1998. Feminist internationalism: Imaging and building global civil society. In *Electronic empires: Global media and local resistance*, ed. Daya Kishan Thussu. London: Arnold, 208–222.

Statistics Canada. 1995. *Women in Canada*. Ottawa: Minister of Industry.

Stern, Jane and Michael. April 15, 1991. Our far-flung correspondents: Neighboring. *The New Yorker*: 78–93.

Stern, Susannah. 1999. Adolescent girl's expression on web home pages: Spirited, sombre and self-conscious sites. *Convergence: The Journal of Research into New Media Technologies*: 22–41.

Stienstra, Deborah. Summer 1996. From Mexico to Beijing: International commitments on women. *Canadian Woman Studies* 16:14–17.

Stoller, Debbie. 1999. Media whores. In *The BUST Guide to the New Girl Order*, ed. Marcelle Karp and Debbie Stoller. New York: Penguin Books, 265–272.

Stone, Allucquere Rosanne. 1995. *The War of Desire and technology at the close of the mechanical age*. Cambridge, MA: MIT Books.

Suchman, Lucy and B. Jordan. 1989. Computerization and women's knowledge. In *Women, work and computerization: Forming new alliances*, ed. K. Tijdens, M. Jennings, I. Wagner, and M. Weggelaar. Amsterdam: Elsevier Science Publishers B.V., 153–160.

Survey reveals Internet shopping habits. January 2000. *Direct Marketing* 62:15.

Sweetman, Caroline, ed. 1999. *Gender and technology*. Oxford: Oxfam.

Tamblyn, Christine. 1994. She loves it, she loves it not: Women and technology. *Proceedings of ISEA '94*, Helsinki, Finland. URL: www.uiah.fi/iseaw/www/artworks/96/96.html

Taylor, Jeanie, Cheris Kramarae, and Maureen Ebben, eds. 1993. *Women, information technology, and scholarship*. Champagne-Urbana: Center for Advanced Study, University of Illinois.

Tech industry nudged to close gender pay gap. May 11, 2000. *CNET News.com*. (Archived at www.news.cnet.com).

Tedeschi, Bob. July 12, 1999. E-commerce report: As women start to use the Internet more for shopping, the prospects sharply improve for on-line retailers. *The New York Times*. (Archived at www.nytimes.com).

The most connected nation? February 17, 1999. Reuters Newswire. URL: www.wired.com/news/business/0,1367,17948,00.htm

Teicholz, Nina. October 22, 1998. Women want it all, and it's all on line. *The New York Times*. (Archived at www.nytimes.com).

Tichi, Cecilia. 1991. *Electronic hearth: Creating an American television culture*. NY: Oxford University Press.

Tierney, John. December 17, 1998. The big city: Women ease into mastery of cyberspace. *The New York Times*. (Archived at www.nytimes.com).

Time Digital. January 29, 1999. The war over women on the web. URL: www.pathfinder.com/time/digital/daily/0,2822,19063,00.html

Transcending the gender information divide—Action plan. March 2000. Women's Forum of Global Knowledge II, Malaysia. URL: www.globalknowledge.org.my

Travers, Ann. 2000. *Writing the public in cyberspace: Redefining inclusion on the Net*. New York: Garland Publishing.

Truckenbrod, Joan. 1994. Gender issues in the electronic arts inform the creation of new modes of computing. *Proceedings of ISEA '94*, Helsinki, Finland. URL: www.uiah.fi/bookshop/isea_proc/high&low/j/16.html

Turkle, Sherry. 1997. *Life on the screen identity: Identity in the age of the internet*. New York: Touchstone Books.

Turnipseed, Kate. Winter 1996. Electronic witches: Women activists using e-mail in the former Yugoslavia. *Feminist Collections* 17:22–23. URL: www.library.wisc.edu/libraries/WomensStudies/fcturnip.htm

UNIFEM. 2000. *Progress of the world's women 2000: A new biennial report*. New York: UNIFEM. URL: www.unifem.undp.org/progressww/2000/index.html

United Nations. 1996. *Report on the expert workshop on "Global information through computer networking technology in the follow-up to the Fourth World Conference on Women (FWCW)"*. URL: gopher://gopher.un.org/00/sec/dpcsd/daw/REP

The United Nations Commission on Science and Technology for Development (UNCSTD). 1997. *Report of the Working Group on Information and Communication Technologies for Development*. Prepared for the Third Session, 12 May. Geneva, Switzerland. E/CN.16/1997/4, 7 March.

United Nations Development Program. 2000. *Human development report 2000*. URL: www.undp.org/hdr2000/home.html

United States Advisory Council on the National Information Infrastructure. 1996. *A nation of opportunity: Final report*. URL: www.benton.org/KickStart/nation.home.html

United States Department of Commerce, National Telecommunications and Information Administration. 1999. *Falling through the net: Defining the digital divide*. URL: www.ntia.doc.gov/ntiahome/ftt99/contents.html

Vered, Karen Orr. 1998. Beyond Barbie: Fashioning a market in interactive electronic games for girls. In *Millennium girls: Today's girls around the world*, ed. Sherrie A. Inness. Lanham, MD: Rowman & Littlefield Publishers, Inc., 169–191.

VNS Matrix and Virginia Barratt interviewed by Bernadette Flynn. 1994. *Continuum: The Australian Journal of Media and Culture* 8:419–432.

Wajcman, Judy. 1991. *Feminism confronts technology*. Cambridge, England: Polity Press.

Wakeford, Nina. 1999. Gender and landscapes of computing in an Internet cafe. In *Virtual geographies: Bodies, space and relations*, ed. Mike Crang, Phil Crang, and Jon May. London: Routledge, 178–201.

Ward, Kathryn B. and Jean Larson Pyle. 2000. Gender, industrializaion, transnational corporations and development: An overview of trends and patterns. In *From modernization to globalization: Perspectives on development and social change*, ed J. Timmon Roberts and Amy Hite. Oxford: Blackwell Publishers, 306–327.

Wertheim, Margaret. 1999. *The pearly gates of cyberspace: A history of space from Dante to the Internet*. New York: W.W. Norton.

Whaley, Patti. 1996. Potential contributions of information technologies to human rights. *Women and Performance: A Journal of Feminist Theory*, Issue 17:225–232.

Wilding, Faith. 1998a. Notes on the political condition of cyberspace. *Art Journal* 57 (Summer):46.

Wilding, Faith. 1998b. Where is feminism in cyberfeminism? posted on the Old Boys Network, www.obn.org/cfundef/faith_def.html

Wilhelm, Anthony G. 2000. *Democracy in the digital age: Challenges to political life in cyberspace*. New York: Routledge.

Winner, Langdon. Summer 1993. Upon opening the black box and finding it empty: Social constructivism and the philosophy of technology. *Science, Technology, and Human Values* 18:362–378.

Winter, Debra and Chuck Huff. Spring 1996. Adapting the Internet: Comments from a women-only electronic forum. *The American Sociologist*: 30–54.

WomenAction 2000. 1999. *Report of on-line discussion on women and media, held November 8–December 17, 1999*. URL: www.womenaction.org/global/wmrep.html

WomenAction 2000. 2000. *Alternative assessment of women and media based on NGO reviews of Section J, Beijing Platform for Action*. URL: www.womenaction.org/csw44/altrepeng.htm

Women.com makes deal to launch British venture. June 22, 2000. *The Globe and Mail Online*. (Archived at www.globeandmail.com).

Women download savings, not software. April 12, 1999. *Brandweek* 40:44.

Women go to the Internet to set up microcredit program. May 10, 2000. *CNET News.com*. (Archived at www.news.cnet.com).

Women, the information revolution and the Beijing conference. 1996. *Women2000* Issue No. 1, October. United Nations Division for the Advancement of Women (DAW). URL: www.un.org/dpcsd/daw/w2ww.htm

WomenWatch Homepage. URL: www.un.org/dpcsd/daw/dawwatch.htm

Wong, Wylie. March 12, 1998. Women will help fuel e-commerce. *Computer Reseller News*. URL: www.crn.com/dailies/weekending031398/mar12dig11.asp

Zeidler, Sue. March 8, 2000. Net shatters not-so-sacred gender myths. *ZDNet News*. (Archived at www.zdnet.com).

INDEX